W9-BPM-881

Beer Money

Beer Money

A MEMOIR OF PRIVILEGE AND LOSS

Frances Stroh

HARPER
An Imprint of HarperCollins*Publishers*

HarperCollins books may be purchased for educational, business, or sales promotional use. For information, please e-mail the Special Markets Department at SPsales@harpercollins.com.

FIRST EDITION

Designed by William Ruoto

Library of Congress Cataloging-in-Publication Data has been applied for.

ISBN: 978-0-06-239315-9

16 17 18 19 20 OV/RRD 10 9 8 7 6 5 4 3 2 1

AUTHOR'S NOTE

This book is my love letter to a past I could reconcile only through the strange alchemy of writing and rewriting its pages over the course of four years. My family's story has been documented elsewhere in less nuanced ways, and I make no claim that my version is the definitive one, though I believe it to be the more felt one. There is no final truth, only one's own; the story on these pages is my story, the one that only I could write. And since the mind naturally conflates events that seem connected, I allowed three scenes to remain as composites because that is how they initially came to me: the events of two family Christmas celebrations are blended into one; two trips to the "Krishna center" in Detroit are also merged; and two gatherings of Taft friends in the apartment on Riverside Drive, in 1983 and 1984, have morphed into a single event. There are no composite characters, though some names have been changed. I reconstructed scenes with dialogue, facts, and details as closely to the actual events as possible, relying on research, journals, old video interviews, as well as recent interviews with several family members. And of course I called on my own memory. Throughout these pages I have been deeply committed to the emotional truth of the story, reconstructing dialogue as I remember it but also allowing the characters to be themselves and to say the sort of things they would have said. Many people and events have been omitted, allowing the part to stand in for the whole and permitting the story to unwind according to its own logic, as it shaped me.

For Mishka

"How did you go bankrupt?" Bill asked.
"Two ways," Mike said. "Gradually and then suddenly."

—ERNEST HEMINGWAY, *The Sun Also Rises*

Contents

Beer Money

THE ERIC STROH FAMILY, 1971

(by Eric Stroh)

PROLOGUE

I stood in the center of a white room, darkened and soundproofed, with six videos playing on screens surrounding me, each one featuring an enormous talking mouth. I was trying to see the installation as others saw it. Visitors to the exhibition took in the cacophony of voices—a family of six telling the family story from disparate points of view.

As I stepped closer to each screen, I could single out its distinctive voice.

"There were years and years during which our family lived in denial of who and what we really were," said one young man.

"I'm sure there are families with problems similar to our own," offered another, "somewhere out there . . ."

"My life has not unfolded as I thought it would," the mother said. "I lived in a dream when I was young."

"Happiness is not something you turn on and off like running water." The father smiled and puffed on his pipe. "Life doesn't work that way."

Over and over again, I had listened to these pronouncements in a cramped, dark editing room. I was a twenty-six-year-old installation artist whose work had been selected for this group exhibition at San Francisco Camerawork Gallery, *The Family Seen*. The ruined family projected on the screens was my own.

The people around me gazed at the talking mouths as if hypnotized, unaware that one member of this family stood in their midst. No one knew they were watching me reflect on my own father's alcoholism or my brother's drug addiction. Nor could they know we were part of a multigenerational corporate legacy whose products were all-American household names. But, then, we'd all been groomed since birth to never discuss this, least of all in front of a camera.

When my brother Charlie said, "Let's hope the family business picks up so I can take early retirement along with everyone else," viewers might have assumed he was a hard-working employee at, say, a faltering hardware store, not a brewery heir turned convicted drug dealer.

When my own mouth on the screen spoke—"I always felt I was somehow special because I was my father's favorite"—I alone knew that my father had abandoned his dream to be a career photographer in order to join the family business, then rejoiced when I myself took up photography. I also knew this video piece could not have been what he had in mind when he gave me my first Nikon at age sixteen.

It was revenue from Stroh's Beer, Old Milwaukee, Schlitz, and Schaefer that paid for things like that fancy camera. My family had been brewing beer in Detroit since long before the invention of the assembly line and the Model T. We'd survived Prohibition by selling ice cream and malt syrup for home brewing, and then entered the mid-twentieth century with a beer brand emblematic of the American dream itself—Stroh's. Our products came to symbolize at once the American working class and, by way of college kids who drank Schlitz at Nantucket beach parties and served Stroh's on draft at their campus fraternities, the carefree youth of the seventies and the eighties. By the mid-1980s, taglines like "Stroh's Is Spoken Here" and Old Milwaukee's "It Doesn't Get Any Better Than This" were as recognizable as Budweiser's "The King of Beers." But within a handful of years a century-and-a-half-old brewing tradition would be verging on extinction, Stroh's Beer vanishing from the national landscape.

Our family business blazed the trail of economic decline in America. By the late eighties our beer sales were faltering even as our personal tragedies had begun to mount: my brother Charlie's drug conviction, my parents' divorce. By 1993, the time was ripe for my video piece; my immediate family members all agreed to the interviews, the floodgates opened, and the story of a real American family flowed forth.

It had filled with people now, the small installation room. They stood in tight clusters, the ambient light of the screens flickering off their faces. The scent of gallery red wine floated in the air. No one seemed to notice my absorption in the content of the interviews.

I stepped closer to Charlie's screen. "As a kid I used to lie to Dad, and he used to whip me. But the only reason I lied to him was because I was intimidated by him."

I moved toward my brother Bobby's screen. "My father is a very talented, very artistic person, and I think it's a shame he hasn't done more with himself." He frowned through his mustache. "When he's dead and gone I wish there would be more he could leave behind, like gallery shows or what-have-you."

I walked over to my brother Whitney's screen. "Dad is an overbearing, controlling son of a bitch. I've tried to get along with him but he's always a prick to me, so fuck him."

The crowd bristled with tension.

"It's so cathartic," one viewer said to her companion in a hushed voice.

"If you like your catharsis with a side of explosives," he whispered back.

We'd each said things to the camera that we wouldn't want the others to overhear. Now our voices rang out in a nexus of crossed indictment. To anyone else, the effect might have been "cathartic" and "explosive"—but not to me. After all, I had lived it. And our downward spiral unfortunately had a long way to go yet.

Many years would pass before I would come to see that the Stroh's Beer story, my family's story, and the story of the once great city of Detroit were all intertwined, our destinies and histories so enmeshed that in their final days the brewery, the family, and Detroit all tumbled together, a long-eroded cliff falling whole into that inland sea.

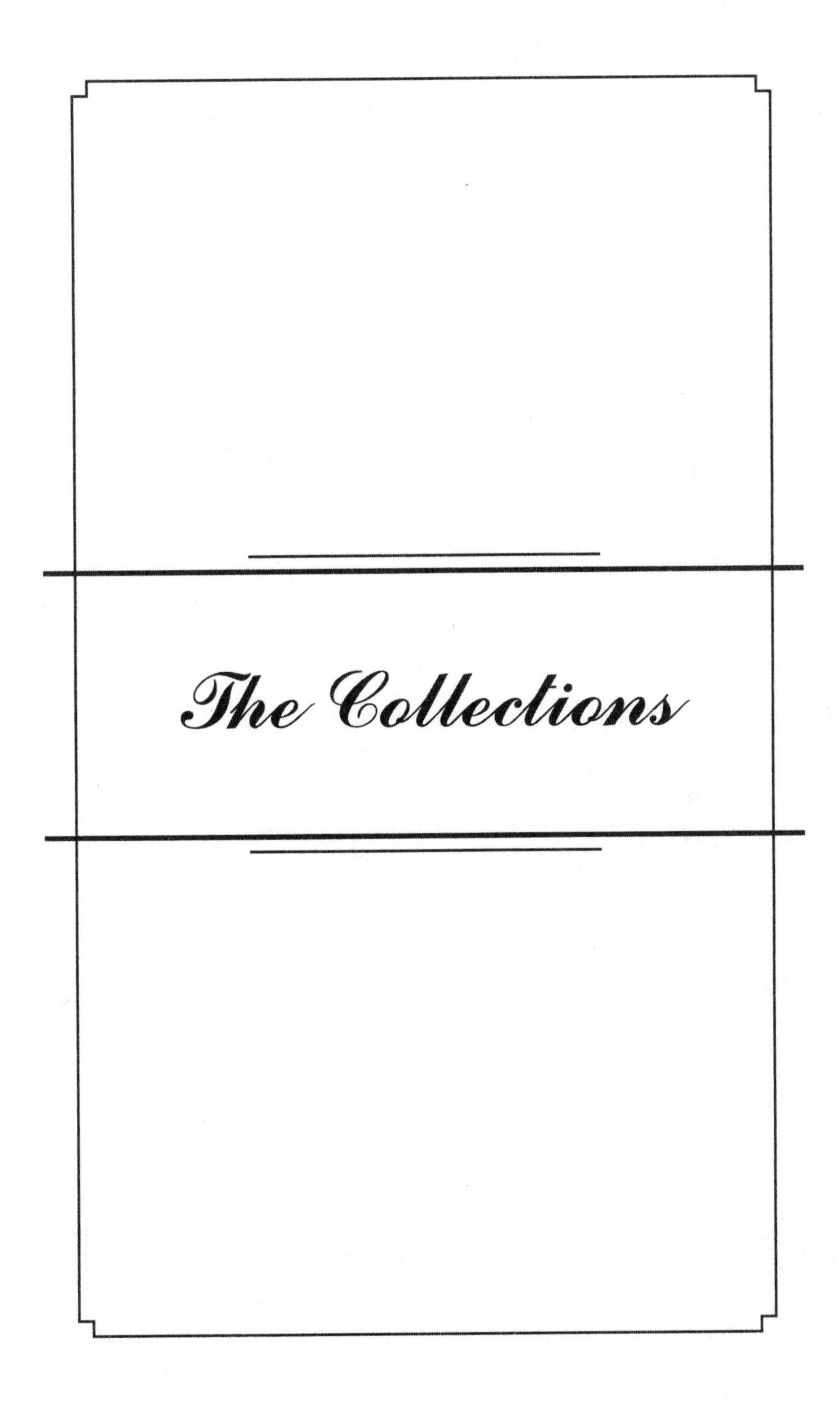

The Collections

FRANCES STROH, 1973

(by Eric Stroh)

New York City, 1973

The shopkeepers of my youth were eager men and women who would turn up the lights and smile extravagantly when my father and I walked though their doors.

"Hello, Mr. Stroh" echoed through the dusty chambers of Madison Avenue antique stores, Broadway camera dealers, the elite pipe shops of London. My father always addressed the shop owners by their first names, as if they were old friends, while they led us toward elegantly appointed back rooms. Armored doors opened into velvety interiors with the most prized discoveries of the season: a pair of nineteenth-century celestial and terrestrial globes; an engraved gold-and-silver-plated antique firearm set. My father handled the items with confidence and familiarity, a cigarette hanging off the corner of his lip, the muted excitement in the room making my breath uneven.

I remember one of our trips to Madison Avenue, when I was six. My father and I stood in the back of the shop inspecting a pair of ivory-inlaid revolvers. The dust in the room spun around in a beam of light from the window. The shop owner broke the silence with chatter, perhaps trying to ease any guilt he imagined my father might feel at spending such excessive sums.

My father interrupted him. "Are the holsters original? I bought fakes in Chicago once."

"Those are double loop Western holsters, Mr. Stroh, by A. M. Nash. A hundred percent antique."

My father ran his thumb over the embossed leather, scratching the finish with his thumbnail. "Patina's good," he said.

The shop owner nodded. "You won't find a better set." His few strands of hair, combed over an expanse of receding hairline, had been carefully gelled into place. He walked over to a dusty shelf of books and pulled out a volume. "Oh, and I meant to call you about this Dickens series, Mr. Stroh." He opened the gold-embossed leather-bound book. "It's a first-edition collection of the totality of Dickens's novels, circa 1880. Twenty-nine in all."

My father put the holsters down on a table and took up the volume. "*A Christmas Carol!*" he said, smiling over at me.

I leaned against a silk-upholstered divan and dug my hands deeper into the pockets of my Sunday overcoat, a vague feeling of dread constricting my throat. We had watched *A Christmas Carol* three Christmases in a row in the library, our tree sparkling in the corner of the living room, glasses of eggnog sitting on coasters next to us. I watched the movie only to please my father. The truth was, I always had terrible nightmares afterward, my sleep haunted by the three ghosts.

As he looked through the volumes, my father's face succumbed to that expression I'd seen so often on these shopping trips, the sort of glow I'd seen on worshippers' faces at Christ Church as they sang from the hymnbooks, and I knew my father would buy the Dickens set no matter the cost, putting him in a good mood for at least a few hours or, if I was lucky, the entire day.

Afterward, my father and I walked over to Park Avenue in the brisk April wind to have lunch at the Regency Hotel, my father's English leather shoes grinding the pavement as I struggled to keep up with him. He took my hand and led me across the street against a throng of rushing people. My father pointed up at one of the tall buildings. "That one there?" he told me. "It's the most expensive apartment building in the world."

The limestone building rose up toward the sky, as if primed to launch, with perhaps a thousand windows. Power surged from its interior, as if the game of the world—a monstrous game of Monopoly always under way—were being played day and night by its inhabitants. The air around us seemed to crackle with excitement, and I felt at once proud that my father knew such things and disappointed to be only passing by. I wished we could go inside the building's walls of polished stone, ride the elevator to the top floor as we had at the Empire State Building, and look out on the world.

"Maybe you'll even live there someday," said my father, sensing my fascination.

I felt thrilled. And doubtful. Though my father constantly bought expensive things, my mother often worried that we couldn't afford them. "Really?" I asked tentatively. "I could live there?"

"Sure." My father smiled. "You can do anything you *want.* All it takes is money."

I could feel the solid warmth of my father's hand as he led me along the sidewalk to lunch; his happy mood seemed to promise me the world.

We walked through the marble-floored lobby of the Regency Hotel into the dining room and were seated and given menus. Plates traveled past us to other tables in the hands of black-tied waiters. My stomach knotted with hunger. The menu looked like the one at our country club—lamb chops, filet mignon, whitefish. My father glanced around for the waiter.

"I'll have a hamburger, *extra* well done, please," he said when the waiter arrived. The waiter raised an eyebrow. "That's right, cook it like a hockey puck."

My father turned to me. "My daughter would like a hot dog, French fries, and a Coke." With the exception of the Coca Cola, the items he ordered were not on the menu.

The waiter came by with a tray of rolls and placed one on each of our plates with a set of silver tongs. Perfect balls of butter sat atop crushed ice in a silver bowl.

"How about we swing by Schwarz after lunch?" my father asked me. He took a roll and slathered it with butter. "Then we'll have tea at the Plaza." His light-blue eyes fixed on mine, and I felt a rush of excitement. He'd remembered.

We'd been to FAO Schwarz—my favorite destination—a few times before. Just down from the Plaza Hotel, the toy-store windows sparkled with elaborate displays, beckoning to every child passing by. One year an entire kingdom of Madame Alexander dolls inhabited castles and locked towers, fought dragons and rescued princesses. "Okay," I said shyly, not wanting to show my father how much I'd hoped he would suggest it.

My father took a second roll, this time from my plate, and smiled at me. Whenever he was happy, I felt I was at the center of a benevolent universe.

Having thrived in Detroit for five generations, my father's clan was infamous for spending money nearly as quickly as they made it, my father's generation in particular.

My great-great-grandfather, Bernhard Stroh, had come over from Kirn, Germany, in 1848 with a family recipe. In 1850 he established the Lion Brewing Company in Detroit because the local water tasted so good. Bernhard made a Bohemian-style brew in his basement and sold the barrels door-to-door out of a wheelbarrow, saving every spare penny to buy a horse-drawn carriage. Later, thanks in no small part to Henry Ford and his Model T trucks, Bernhard's sons, Julius and Bernhard Jr., expanded the company's distribution throughout the entire Midwest, renaming it the Stroh Brewing Company.

By the 1970s, the third and fourth generation of Strohs were running the family-owned brewery. They made a regional beer brand—Stroh's Beer—that went national in the early 1980s after the purchase of the Schlitz and Schaefer breweries, a consolidation of the industry that landed thirty beer brands in our portfolio, making the family company the third-largest beer maker in the United States, behind only Anheuser-Busch and Miller. The majority of Stroh's brands targeted inner-city subcultures, the blue-collar segment, and—because the beer was cheap—college kids. At its peak, the Stroh Brewing Company launched an enormous commercial and residential real estate project in downtown Detroit, built its own biotechnology research center in Durham, North Carolina, and underwrote a private plane for its CEO. Named in the Forbes 400 list from 1984 to 1992, the Stroh family possessed the largest private beer fortune in America.

For decades, the money was flowing and the Strohs lived like kings. My father's notorious collecting landed him on every dealer's A-list, making him the poster boy for the Strohs' spending habits. He loved the attention, the grandiosity, and the elusive hit of immortality he felt when he walked into a shop. It seemed inexhaustible, the pipeline of beautiful objects—and the money to buy them—and we never grew tired of wandering the shops' dusty back rooms.

But as my father's health declined in the decades to come through the various stages of heart disease, and my life and work took me elsewhere, our shopping trips gave way to brief visits in this or that city, to catch-up calls with tenuous overseas connections, and the team we'd formed in my youth slowly dissolved.

My father died alone in the hospital in 2009. I was stunned he hadn't wanted me at his side. He hated to show vulnerability, of course; still, it hurt that he'd been so stubborn all the way to the end. In my sorrow I realized that the small girl who so loved and admired him had never really left. It was that same small girl who despairingly called his answering machine in Michigan for months, until the house was sold, just to hear his voice on the outgoing message, incredulous that he was no longer there.

When the time came for me, as the executor of the estate, to put my father's collections up for sale, a crippling fatigue settled in; I yearned to wade in my grief for as long as my spirit needed to, not haggle over consignment agreements and auction contracts. My father had left me the whole of his collections—a nod to the years we'd spent together buying them, and perhaps, as some sort of apology. The gesture, though, was like a loaded pistol; the Stroh Brewing Company had been sold ten years before, in 1999, and my father had spent the bulk of his share of the proceeds. The collections—and the Grosse Pointe house in which they sat—were all that remained of my father's legacy, and disinheriting my brothers seemed nothing less than cruel.

It felt as if the collections and the money they represented had formed an invisible web in which I'd been caught all my life, and I found myself secretly wishing I could give everything to charity with a single phone call. But as I'd decided to split my father's possessions with my brothers, or at least their value, assessing, dividing, and selling off the collections was what I had to do. The freedom I'd coveted came only gradually, as things of real value so often do.

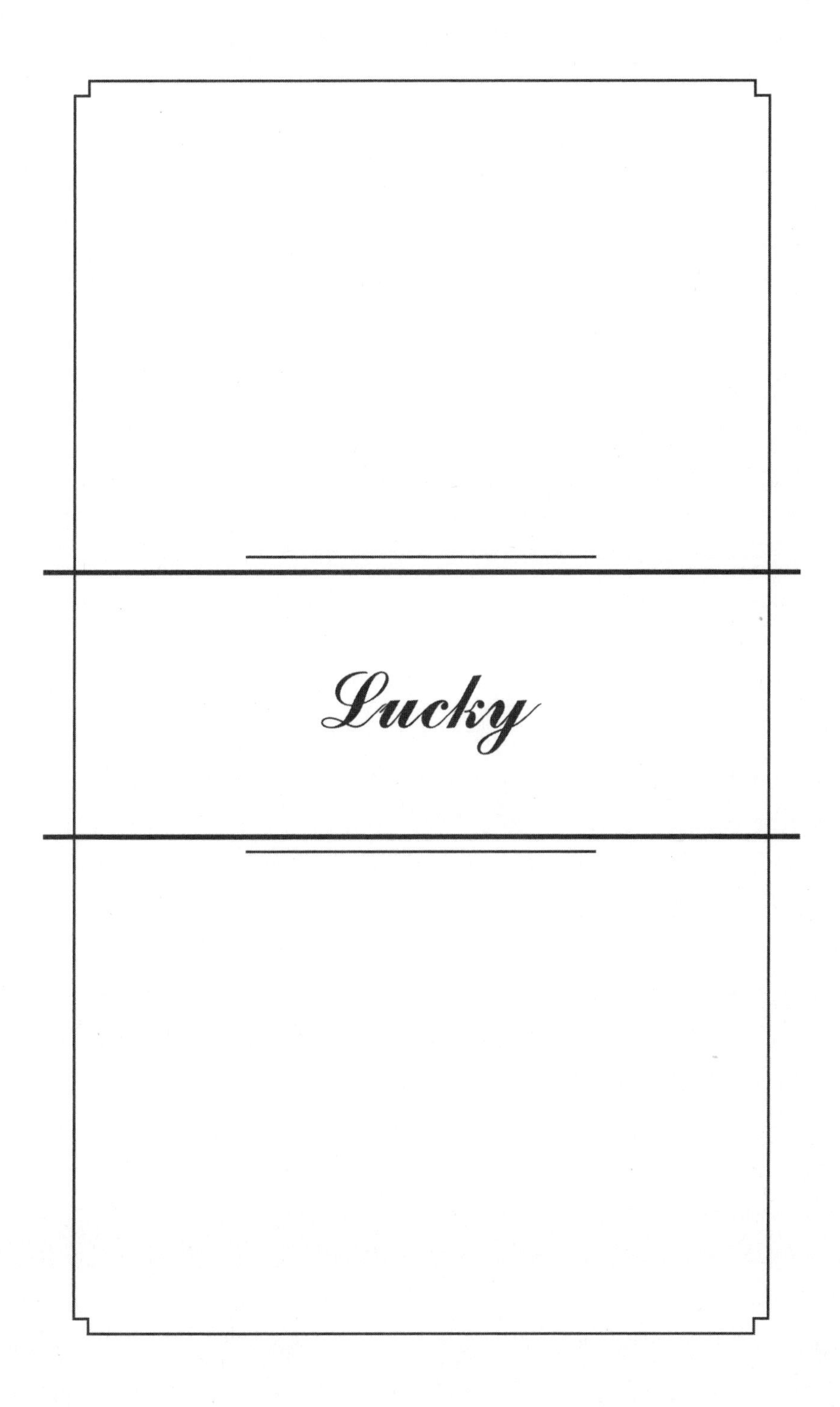

Lucky

STROH BREWERY

(Copyright 1973 The Detroit News, All Rights Reserved)

Detroit, 1973

"Lock your doors, kids," my father said as we crossed into Detroit on Jefferson Avenue, leaving behind the wide green lawns and lakefront mansions of Grosse Pointe.

Dropping their *MAD* magazines on the backseat, Bobby and Charlie sat up at attention. I hugged my Barbie to my chest. We had entered the fear zone. Miles and miles of derelict buildings stretched before us, four-story prewar brick buildings with boarded-up windows, peeling advertisements, and torn awnings. Many of the structures looked as if they once had been rather grand houses or apartment buildings, their graceful stone steps rising up to paneled, arched wooden doors. I imagined women in wide feathered hats coming out of those doors, their uniformed drivers waiting outside in horse-drawn carriages. Now the buildings' brick walls were collapsing.

We pushed down the plastic knobs at the tops of our doors, listening for that reassuring *click* of safety, and sat silently for the remainder of the drive to the brewery, as if being quiet might attract less attention to our father's silver Chrysler.

"We're only as safe as the locks on our doors," my father always said.

We knew why—because all the people on the street were black. Men and women walked into liquor stores that had the word *LOTTO* spray-painted on their awnings, cradling brown paper bags when they came out. Cadillacs crawled along side streets where the houses had been burned in the riots and left to disintegrate. Women wore short skirts in November, their legs muscular and lean above high heels. Men circled in the middle of the road, back and forth, back and forth—angry, wild-eyed, shouting at each other.

I wasn't as afraid as my father. Sometimes, when my parents traveled with their friends to Bermuda or the Bahamas, I stayed in a black neighborhood in Detroit with our housekeeper, Ollie. I climbed trees with Tony and Dana, Ollie's grandchildren, and sang gospel at her church. I ate Ollie's Southern cooking and watched her husband, Raymond, blow cigarette smoke out of his tracheotomy hole. Raymond was dying of lung cancer. I could hear Ollie crying at night through the paper-thin wall and Raymond comforting her by humming old songs, and I wondered why, when it was perfectly safe to stay at Ollie's, we had to lock our doors to keep the black people on the street out.

"Damn riots," my father said. "Changed everything. We could hear the gunshots and smell the smoke all the way up in Grosse Pointe."

They had come the year after I was born, the riots, in

1967. My father said the blacks had changed after that, but of course he wasn't talking about Ollie. The whites fled Detroit for the suburbs, and the Grosse Pointe police force doubled in size.

"Any nonresident black found within the city limits will be escorted back to Detroit," I'd once heard a police officer say to a woman who'd complained about black kids swimming at the Farms Pier pool. She'd been dressed in a monogrammed pink-and-green sweater, and her husband's khakis were cuffed at the ankle, like my father's. They looked like everyone else in Grosse Pointe, the kind of people who drank cocktails from glasses etched with the motto "You can't be too rich or too thin," and whose black cooks and maids were treated entirely differently from the blacks on the streets or at the parks.

"Whites aren't safe down here anymore," my father said, switching lanes to avoid hitting a man who carried a boom box on his shoulder. "Coleman Young's made sure enough of that."

Coleman Young was the mayor of Detroit. My uncles, who ran the family company, were always having meetings with "Mr. Young," making "deals" with him. Uncle Peter and Great Uncle John ran the brewery, Uncle Gari the ice cream division. As children, we understood Coleman Young to be the king of Detroit, someone our family had to please at all costs, because we were white.

I could see the Stroh's Beer sign just ahead, hovering above the brewery in red block letters that lit up the sky. It always startled me, seeing our name like that, and I looked away as we turned into the parking lot, focusing instead on the rows and rows of blood-red beer trucks, *Stroh's Beer* inscribed in gold across their sides.

My father swung open the door to the Brewhaus, allowing Bobby and Charlie through. My father smiled down at me as I passed through in my red winter coat, one of my mittens trailing on the floor from the string connecting them through my sleeves.

All through the cavernous space was the pungent scent of hops and wheat. Enormous copper cauldrons of brew, one after another, emitted their noxious steam as we walked a catwalk running along the perimeter of the space. We looked down at the blue-uniformed men adding ingredients to the brew through sliding hatches on the sides of the cauldrons, their rosy copper gleaming under the fluorescent lighting.

"Can you smell that beer?" my father shouted over the din of machinery. "It's cooked with real fire." He pointed to a row of six copper cauldrons that had been tiled around their sides, like bathtubs. "The fire's inside."

I was just learning to read. I remembered seeing the words *Fire-Brewed* on a beer bottle in our refrigerator.

"Only fire-brewed beer in the U.S.," my father told us as he stamped out his cigarette on the catwalk. "We do it the old-fashioned way."

My father worked in the marketing department. Sometimes he flew to Hollywood to oversee the production of Stroh's Beer commercials. Later, he'd show me the ads on TV while we sat eating pizza in the library at home. My favorite was an ad in which a pretzel climbed up a bottle of beer to take a sip.

My father led us around the perimeter of the Brewhaus. He tapped his cigarette pack on his open palm to knock one out. He wore a dark-gray pinstriped suit with a white shirt and a burgundy tie dotted with tiny tennis racquets. He had on businessman shoes—lace-up black barges with pointy toes and tiny eyelets in the hard leather, shoes that weighed as much as a small dog when I picked them up in his closet, especially with the shoe horns still in them.

"Dad, wear *these*," I would say as he dressed for work, holding up the leaden shoes. He had at least ten pairs to choose from.

He always walked over in his black socks, held up by suspenders just below his knees, and took the shoes. "Thanks, Minuscule."

My father leaned into the railing of the catwalk. He looked at Bobby and Charlie, as he drew thoughtfully on his cigarette. "You two will work here someday. This is your company."

"I know, Dad," said Charlie, as if he'd heard my father say this many times.

Bobby and Charlie studied the men who loaded the hops into the vats. Bobby brushed his auburn curls out of his eyes. At twelve and fourteen, my brothers were big boys and, with their tweed jackets and corduroys, seemed nearly ready to don their own business suits. Fair skinned and freckled, like my mother, Bobby went to a boarding school in Connecticut called Kent—the same name as my father's cigarette brand. Charlie, too, would soon go away to school. He shared my father's coloring—straight blond hair, blue eyes, rosy skin—

none of which stopped my father from favoring Bobby, his firstborn.

"Dad, can I have *that* job?" Charlie asked excitedly, pointing down at a man who took the temperature of a glass of golden liquid.

"Sure, Chas," said my father, "maybe some summer when you're in college."

Where would I work at the company? I wondered. I'd seen only one woman since we'd arrived—at the reception desk. "What about me, Dad?"

My father smiled his Hollywood smile. "You? You're going to be a movie star, right, Franny?"

This was one of our inside jokes. My father adored old movies—anything with Humphrey Bogart, Fred Astaire, Bing Crosby—and their beautiful leading ladies: Lauren Bacall, Ingrid Bergman, Audrey Hepburn.

We crossed into the bottle shop, where an endless procession of dark-brown bottles were filled with beer, labeled, and sorted as they traveled through a mechanized assembly line—a miniature of the one I'd seen at the Ford plant on a school field trip. I watched, fascinated, as the bottles marched along, like ants, toward some mysterious place where they would be loaded into cases and trucked away.

"My father built this packaging facility just before he died, in the late forties," my father said with pride. "Should have put in a can plant, too, of course. That had to be added later."

These were some of the only facts he ever shared about my grandfather, who'd died of lung cancer in 1950 when my father was seventeen. Sometimes I wondered if my father had even really known him.

Bobby's freckles glistened in the rising heat. He pulled at his shirt collar. "I want to sell beer, Dad," he said, turning to our father. "Can I?"

"Sure, why not?" My father turned and walked us around the length of the bottle shop. I tripped in my winter boots, and Charlie took my hand. Bobby removed his jacket and swung it over his shoulder. He wore a brass Stroh's belt buckle.

My father unlocked a door that opened onto a long corridor lined with portraits of our beer-making ancestors and their wives—Bernhard, Eleanora, Julius, Hetty. Some of the portraits had been painted by Gari Melchers, a well-regarded American Impressionist who also happened to be my great-great uncle.

My father stopped in front of a painting of Julius Stroh, his grandfather, who gazed quizzically at us through a monocle, severe and determined in his morning coat and cravat.

My father smiled up at the portrait. "I used to sit on Julius's lap and he would always say, 'Have you been a good boy today, Eric?' and I knew I had to lie. 'Yes, Grandpoppy, I have.'" My father laughed. "The old kraut could be awfully punitive. So could my father, for that matter."

He flicked open a gold Dunhill lighter, one I recognized from our trip to London the previous summer, and lit his cigarette. He bought the lighter the same day he took me to the Tower of London. I remembered him pointing out the medieval torture devices used on the kings' disobedient subjects. The chopping block still had the ax marks where real heads had been severed. Afterward, we'd gone to Harrods, where my father had bought me a pile of new summer dresses. But by the time the dresses arrived home, I had already outgrown them.

"What a *terrible* waste of money," my mother complained. "Eric, how could you be so reckless? This is why I never buy children's clothes new."

My father stood in the dining room, emptied the shipping box, and looked through the dresses, a defeated expression on his face. He had been so excited to see me in them. I'd worn only one blue chiffon dress on the night we'd gone to see *Alice in Wonderland* at the theater.

We all regarded the painting of Julius hanging menacingly over our heads in its heavy gilt frame.

"Would Julius have spanked us?" asked Charlie.

"Would he have *spanked* you? You kids don't know how good you have it," my father said wistfully.

Next he led us down the corridor to a heavy wooden door that opened to the Rathskeller, a light-filled welcome room for brewery tour guests arranged with red-and-white checkered tables and decorated with Stroh memorabilia—antique beer trays with the old Stroh's logo, hurricane lamps, air balloons, and giant, brightly painted toy beer trucks—all with *Stroh's Beer* decaled in gold. I wanted to touch everything, run the trucks across the floor like my brothers used to do at home, but most of it sat on high shelves well out of reach.

We were the only guests. Everyone who worked in the Rathskeller greeted my father. The bartender, a bald man wearing an apron and leaning against the wooden bar, seemed especially friendly to my father, like all the bartenders at the clubs my father frequented.

"Well, *hello*, Eric."

"Hello there, John."

"You've got your brood along with you today."

"Most of 'em, anyway."

Whitney, the baby, had stayed home.

"A good-looking lot, they are," said John.

"Let's have a drink, kids," said my father, heading for the bar.

Bobby, Charlie, and I settled at a table and waited quietly. A waitress brought us cheese, crackers, and three Cokes. No one said any more to us, and I began to feel as if the staff in the Rathskeller were waiting for us to leave; they were expecting a tour group "any minute," I heard John tell my father.

My father came over carrying a pile of T-shirts that said "Stroh a Party!" across the front, and we put them on over our clothes. We sat saying nothing in the too-big T-shirts while my father had another drink. And then, from the factory floor, came the hiss and roar of the flames firing up underneath the copper cauldrons.

Charlie broke into a wide, bucktoothed smile. "Cheers, Franny," he said, tapping my glass with his. He tossed the Coke back in three quick gulps—a perfect imitation of my father.

THE HOUSE AT GRAYTON ROAD, 1974

(by Eric Stroh)

Our house was full of shiny valuables that we were forbidden to touch. Rare Martin guitars leaned against upholstered chair backs in our living room, as if waiting for cocktails to be delivered. Glossy silver boxes housing monogrammed guitar picks littered the mahogany tabletops. Antique Leica cameras and real guns from the Wild West decorated desktops and bookshelves, where leather-bound first editions of Nathaniel Hawthorne comingled with a first edition of *Through the Looking-Glass* signed by Lewis Carroll.

On occasion, my father would take out his favorite antique revolver, a perfect greyhound of a gun with a long barrel and intricate engravings surrounding its ivory-inlaid grip, and show me how to clean and oil it. Then he would have me reach my finger around the trigger and pull. Afterward he'd open it up and show me why it hadn't fired; there were no bullets, see.

"John Wayne carried this gun in *How the West Was Won*,"

he'd tell me with pride. My father loved old Westerns. He sometimes dressed the part and walked around the house in boots, spurs, and a cowboy hat, flipping the guns out of their holsters like the outlaw Jesse James. He often told the story of dressing up as a cowboy when he was a kid on Christmas Day, and how his mother had shouted at him to change into a jacket and a tie. "I never forgave my mother for that," he would say with a faraway look.

If my mother came into the room while my father was showing me how to load the bullets, he'd look up at her and beam. "God, I love my guns, Gail—more than anything in the world."

"Those *awful* guns." This was her habitual reply. "How can you love them?" And while the two of them dueled it out with their scripted conversation, I enjoyed the privilege of handling the goods. My parents often talked this way; I usually knew what one would say to the other, a predictability I found deeply comforting. Nor did it concern me, the thought that my father loved his possessions at least as much as he loved us. I took it as a given.

But there was a double standard in our house. While I was sometimes allowed to handle my father's treasures, as soon as my younger brother, Whitney, could walk, he was punished just about daily for touching my father's things—spanked, yelled at, and sent to his room. None of us was immune; when I was four I wandered into my parents' bathroom and ran my father's razor up my arm to see how it worked. When he found it clogged with hair, he slapped me across the face three times, until I admitted I'd done it.

And yet Whitney had these run-ins with my father more than all the rest of us put together.

One day my father left a fragile clay pipe on a table—well within reach of a toddler's wandering hands—and then snapped like a mousetrap when Whitney of course broke it.

"Why the hell did you break my pipe?" my father angrily demanded, taking Whitney by the shirt collar.

"Because I did," said my frightened three-year-old brother.

"Then I'm going to spank you," said my father.

"Why?" asked Whitney.

"Because I am. That's why."

My father and I often went out to a local diner for dinner, just the two of us, while my mother stayed at home to make dinner for the boys. Other times my father and I played a game that I secretly hated, a game ostensibly designed to teach me a valuable skill: how not to get kidnapped.

I still remember the first time he made me play. "*Frances!*" He shouted from the bottom of the stairs. "Time to practice. Step outside, please."

My whole body tensed at the sound of his voice. I could tell from the sour smell of the air around him that he'd been drinking.

Because we were a known family, because we had a name that made us stand out, it was very important, my father had told me, that I play this game with him. Nothing mattered more than this—not even learning how to swim, or how to read, or how to hit a tennis ball.

"They'll take you away and we'll never see you again," my

father said. "Or they'll ask for a ransom that we can't possibly afford to pay."

"What's a ransom?" I asked.

"A lot of money—you know, millions of dollars."

It seemed a terrible curse to have a recognizable name. But to have a recognizable name and not enough money to pay the ransom?—nothing less than a cruel joke.

We walked outside.

"Stand right here," said my father, his brow unaccountably heavy with anger, pointing to a spot on the sidewalk in front of our house, a six-bedroom Spanish Mediterranean with a sprawling green lawn.

This was a serious game, a game that made my father stern, impatient.

I assumed my position on the pavement, a surge of dread twisting my insides. The facades of neighbors' houses suddenly became menacing faces; the sound of their gardeners mowing, a barrage of violent sound in my head. No one, it seemed, would be able to save me.

"I'll go get the car," said my father, walking across the lawn to the driveway to start the car. Then he drove around the block. Sometime during those two minutes, while I waited alone for the car to reappear from the opposite end of our street, something magical and mortifying happened: my father's silver Chrysler became someone else's car, the car of a complete stranger.

My heart jumped at the sight of the approaching sedan, my abductor's car, as it slowed down upon spotting me, a small blond girl, seven years old, with straight-cut bangs.

"Come here, little girl," a frightening man with my father's face called tauntingly from the car. He held a Hershey's chocolate bar out the window.

I burst into tears and ran away, just as I had been instructed. Kidnappers, I knew, baited young children with candy. But even as I raced up to the front steps of our house, I found no comfort in the familiar. It was the creepy, almost psychotic look on my father's face as he called out to me, pretending to be someone else, that terrified me.

Charlie sometimes hid me in his room when my father arrived home from work. We'd sit on his bed and look at books about World War II that he'd borrowed from Bobby, books with pictures of Germans in belted overcoats saluting a man with an Oliver Hardy mustache.

"That's Hitler," Charlie would tell me. "The one who killed all the Jews."

I had seen a picture of the Jews once, naked together in a room with no windows. Their hollowed-out eyes stared right through me, and I'd had nightmares for days afterward. "Let's not look at that book again," I remember telling him, and he hid the book underneath the bed.

Then he'd get out the Indian beads—thousands of them, sorted by color—and we'd string necklaces. He'd gently teach me how to tie a knot, or combine the beads in a particular pattern, and, if my knot didn't work, he never became angry when my beads fell all over the scratchy wool bedspread. He'd just pick them up, one by one, and drop them back into the storage box. Sometimes we could hear my father shouting at Bobby, or shouting at my mother down in the kitchen, and we'd go on stringing beads as if nothing were happening.

Eventually my father would throw open the door to Charlie's room, and he'd send me downstairs so he could fight it out with the boys. Afraid, I'd go into the living room where my mother was busy reading a book, and we'd silently listen to my father chasing Bobby and Charlie up and down the long hallway upstairs as if trying to corral a particularly willful herd of cattle. Loud thumps echoed through the floorboards after my father managed to catch my brothers, but my mother never so much as looked up from her novel.

It was as if I were experiencing the blows myself, compounded by a terrible sense of helplessness. To distract myself, I'd get out my markers and a sketchpad and draw movie stars or Bible scenes I'd studied in Sunday school. I'd spend hours getting the face of Humphrey Bogart or Jesus Christ just right, starting over again and again until the eyes, nose, and mouth were perfect. At school I was universally known to be the best artist, and at home I was mostly left alone to do my drawings, an invisible bubble forming around me, blocking out everything happening inside the house.

Having formed a close friendship in early childhood, Bobby and Charlie spent most of their time together, with Bobby as the leader. When Bobby built a model of the *Ti-*

tanic, Charlie would also build a model of the *Titanic*—only not as well.

One summer evening when they were eight and ten years old, Bobby and Charlie jumped on their bikes and rode up to Rose Terrace, the Dodge family's lakefront estate, to wait in line for a public viewing of Mrs. Horace Dodge's open casket. Once inside, they filed past her body and, to the shock and dismay of the other spectators, Charlie reached out and pinched the corpse's overly rouged cheek while Bobby giggled approvingly.

Charlie struggled with his schoolwork, often having to retake tests, catch up on reading, and endlessly practice word pronunciation with my mother, and yet he was a favorite among his teachers. In hushed voices my parents sometimes referred to him as "slow," but he never received any educational testing or special attention at school. My parents' expectations were simply lower for him than for Bobby, and they deemed Charlie's approval-seeking behaviors as natural for a child with his lesser intelligence.

My father's nicknames of "Chas" for Charlie and "Nit-Wit" for Whitney made his allegiances clear; among his sons, Bobby was the crowned prince. He laughed at Bobby's jokes, encouraged his rare-beer-can collecting, and praised his term papers. Charlie could only trail in Bobby's shadow, awkwardly adopting his hobbies and witticisms. The gentler of the two older boys, and the lesser student, Charlie suffered the brunt of my father's unpredictable moods, absorbing his scathing criticism and bullying until he escaped to the relative safety of boarding school.

As for my mother's oblivion to our problems, this was contrived, she later told us, to protect. "If I had intervened, Dad would only have been that much harder on all of you."

The youngest of three brothers, my father was known as an eccentric boy who would sit on the street curb smoking cigars and mouthing off to kids who passed by. He'd spend his afternoons alone watching nickel Laurel and Hardy movies at the Punch and Judy Theatre, then go home to shower and change into a coat and tie for dinner. When he arrived, his much older brothers, Gari and Peter, would be sitting in the living room with their mother as she sipped her cocktail, awaiting their father's arrival from the brewery, while the cook prepared the usual four-course dinner in the kitchen. No one ever took much notice of my father as he came into the room, save for when the Brylcreem he used in his hair rubbed off on the sofa upholstery, and his mother would shout, "Get off that sofa, you little pest!"

His father, Gari Stroh, who ran the brewery, was forever preoccupied with problems at work, particularly during World War II, when hops and wheat were being rationed. He refused to water down the family beer, thereby sacrificing quality of taste, as the other U.S. brewers had done to keep volumes up. A man of principle, he also had a fierce temper, and my father rarely talked about him. "I was intimidated by my father," he once told me, "and avoided him as best I could."

My grandfather's sternness may have stemmed in part from his guilt over a terrible injury he'd caused his younger brother, John, when they were children. One July afternoon the two boys played in the garden of their Italianate mansion on the Detroit River, taking turns at target practice with a new archery set, while their nannies took tea in the shade of the terrace. Julius, their father, was holding court at a brewery board meeting; Hettie, their mother, was taking a nap before dinner. My grandfather loaded the bow, testing its flexibility, while eight-year-old Uncle John disappeared behind the tree to fetch the stray arrows. Gari aimed at the target, but his finger slipped and the arrow shot sideways, bouncing off a tree's trunk and piercing young John's right eye. The bloody scene that followed, combined with the wrath of Julius, became as legendary as it had been traumatic. For the remainder of his life, John wore a glass eye in his right socket, while Gari wore a sheath of self-recrimination.

My father mostly went unnoticed as a child. When he wasn't alone at the movies, he spent his time with a nanny while his mother and her sister, Louise, enjoyed martini lunches at the country club. His parents, however, still tried to exert control over his activities. My father loved country and bluegrass music, for instance, but was forbidden by his parents to listen to "hillbilly music" at home. At least the driver who brought him to school allowed my father to play country music in the car.

Photography, my father's second great passion, was discouraged as well. After being accepted to the Rhode Island School of Design at age eighteen, my father sought his mother's advice.

"If you go there," she told him, "I'll always feel ashamed. *Can't* you find another college?"

Susie—she insisted her grandchildren call her "Susie," so as not to feel old—had never overcome the embarrassment she felt growing up in the midst of Philadelphia society as the daughter of an artist. Her father, my great-grandfather Nunk, a kind-natured antique dealer and hobbyist sketch artist, often painted the furniture in his shop with early American floral motifs. Susie had always aspired to more—to wealth, glamour, social position.

In spite of the shame she felt over her father's vocation, Susie was Nunk's favorite daughter, not to mention his most beautiful. With her chiseled features and slim figure, she was the daughter for whom he waited up at night with sandwiches and hot chocolate when she arrived home from parties. When the family could afford only one new dress, the other daughters, Louise and Betty, were inevitably passed over in favor of Susie.

One weekend at a friend's country house, Susie met the somber and serious Gari Stroh, son of the renowned brewer Julius. Susie's liveliness and beauty captured Gari's attention, and the two were married within a year. Nine years later, Gari's younger brother, John, married Susie's younger sister, Louise. No longer did anyone need to fret over who got the new dress.

Turning down his acceptance to the Rhode Island School of Design, my father attended Michigan State College instead, dropped out after two years, and then, following a stint in the army, joined the family business. Photography, for the remainder of his life, was relegated to a hobby, while each morning he grudgingly put on a suit and drove to Detroit to the brewing company his family had owned and operated for four generations.

One summer evening, my cousins Pierre and Freddy came over with their parents, my uncle Peter and aunt Nicole. Pierre and I were best friends, born only ten days apart; Freddy was two years younger. We sat in the fading sunlight on the terrace, breathing in a garden bursting with bloom and the raw scent of freshly cut grass, and drank Vernors ginger ale while the grown-ups had cocktails and argued about the brewery. Unlike my parents, Aunt Nicole and Uncle Peter never argued when I was at their house, where I spent many a weekend in the summers, so I knew it must have been my father who started the fight. Pierre and I giggled uncomfortably every time his father made my father raise his voice. We didn't understand what they were talking about, but we knew his father would always win the argument.

"Times are changing, Eric," Uncle Peter gently instructed my father. "We have to grow. Have to get bigger to compete, or we'll continue to lose market share and volume."

My father frowned through a cloud of cigarette smoke. "*How* big?" he asked his older brother, his striped dress shirt dampening under the arms.

Uncle Peter had once had plans to join the CIA, but he had been run over by a truck and nearly crippled, ruining his chances. Now, walking with a slight limp, he ran the brewery alongside my great-uncle John, who, after Gari died, watered down the beer formula to suit the post–World War II American preference for milder beer. After that, the company had grown rapidly, acquiring its number one competitor in Michigan, the Goebel Brewing Company.

"As big as we can *get*," Uncle Peter replied, his jaw set. "Stroh's and Goebel alone aren't enough. We have to keep acquiring other brands, other breweries."

"We need national recognition, Eric," said Aunt Nicole, in her French accent, her chin-length blond hair pulled back with tortoiseshell combs. She was always chic and sophisticated like that, a *Town & Country* magazine cover come to life.

Our housekeeper, Ollie, and my mother arrived from the kitchen with a hot cheese dip and a platter of Triscuits, and Ollie put the dip down on the table with an oven glove. It was hardly like at Aunt Nicole and Uncle Peter's house, far grander than ours, where servants brought out French cheeses, bowls of olives, and delicious sliced tomatoes from their garden, drizzled with olive oil.

"Thank you, Ollie," said Uncle Peter. He always stayed in the kitchen for a few minutes and talked with Ollie, whenever he came over to the house, asking after her family.

"Mr. Peter Stroh is *such* a nice man," Ollie always said after he'd left, and my father wouldn't say anything.

But he *was* a good man. Once, after a golfing accident in his garden when a driver swung into my head, Uncle Peter had driven me to the hospital. All the way to the emergency room he'd held my hand as I'd bled into a linen towel. "You're going to be okay, Franny," he said over and over, his small eyes dewy and bloodshot, and I could see how shaken he was.

The leaf-dappled light had drifted from the terrace to the lawn while the grown-ups talked, and the mosquitoes were starting to nip at my ankles. Finally, my father stood up from the table in a fit of frustration. "And where the hell are we going to get the *money* for all this growth? Our volumes are already starting to decline."

"Exactly," said Aunt Nicole. "The competition is gaining ground; we can't defend our position in the Midwest anymore." She crossed her bone-thin legs, her gold sandals offsetting her tan to perfection.

"We have to grow or go," said Uncle Peter with finality. "We'll borrow the money."

Everyone fell silent. I didn't know what "grow or go" meant, but I understood that the company was in trouble, and I felt afraid. Even Pierre had stopped laughing. The acid smell of the hot cheese dip hung on the air. My father walked away, defeated and angry, like the champion runner I'd seen on TV who'd gotten second place in the Moscow Olympics. He walked the garden's circuit, taking quick drags on his cigarette, absently checking the sprinkler heads. He never liked to sit with any group of people for very long, especially after an argument, and I had that sinking feeling I so often got when my father was unhappy. The evening continued without him, the sun setting behind the trees in bursts of pink and orange, ice clinking in people's glasses as the rest of the grown-ups managed to talk on amiably enough.

Taking my picture was one of the few things in life that made my father happy.

He'd get me to pose against the textured bark of a tree, or the long grass in a field, zooming in close with his vintage Leica, his cigarette dangling as he barked out art direction: "Relax your hand on your knee! That's right. . . . Now, on the count of three." Invariably he'd snap the shutter on the count of two.

Once he'd photographed me in our garden just before a party. I stood beneath a towering oak tree, its gloriously gnarled roots dwarfing my tiny feet. I remembered how my feet ached gripping those roots as I crouched down for the shot.

"This time give me a smile," my father said.

I shifted my weight onto my other leg, my back scraping against the tree. "Ow!" I whined.

My father stood and stamped out his cigarette on the perfectly manicured grass. Sometimes I'd see the gardeners picking up cigarette butts on their way to the flowerbeds.

I tried to stand up.

"Stay there!" my father commanded. "Just a few more."

He breathed heavily as he looked through the viewfinder. He always tried several angles, smiling at me between each one, while my shins throbbed and my feet went numb. I remember wondering if one day I would also learn to use a complicated camera like his Leica. Like my father, I wanted to freeze people in time, as if by doing so I might come to understand them better.

"Frances!" my father said then affectionately. "You look marvelous!"

I smiled for the shot and he clicked the shutter.

My father's photographs were widely recognized as the best in Grosse Pointe. An amateur with little training, he shot the portraits that my parents' friends sent out in their Christmas cards and framed on their living room side tables, portraits that in years to come their children would take into their more modest houses as keepsakes of an all but bygone era of lakefront estates, Lilly Pulitzer dresses, and big, strapping American cars.

Money was everything in Grosse Pointe. You couldn't live there if you didn't have it, and some had a lot more than others. The social hierarchy favored the richest, oldest families who had settled in the area and built Detroit from the ground up—the same families who lived on my street, whose children attended my private school, and who swam at my club. These were my childhood friends, progeny of the Fords, the Fruehaufs, the Chryslers. And despite the age-old taboo, my friends sometimes discussed their families' wealth.

"We have sixty million dollars!" Alison Goodyear announced poolside at the club one day.

In my nine-year-old mind, I tried to picture $60 million: piles of stacked bills as tall and wide as the Goodyears' enormous house. I was already aware that if you were rich, you could be as bad as you wanted and get away with it. My parents lately had been gossiping about Alison's parents, who were getting a divorce, and about the many boyfriends her mother had been spotted driving around in her new Jaguar convertible. I never saw Alison's parents when I had sleepovers at her house. We were attended instead by a multitude of nannies and housekeepers. The parents, I imagined, were off pirouetting from the wrecks of sleek sports cars, martini glasses still in hand.

When I told my mother about Alison's pile of money, she sighed woefully. "That is simply terrible," she said. "Children should not be talking about money."

But money was a part of one's identity. If you lost it, you lost yourself, or so it seemed to me, and throughout my young life I was keenly aware that this is just what we were doing—losing ourselves. "Shirtsleeves to shirtsleeves in three generations," as the saying goes. Even as children, surrounded by so much abundance, we had been warned there wouldn't be enough to last.

"One day, the money will go," my mother often told my brothers and me. "Especially if Dad continues to spend this way."

Her warnings encroached like a black cloud on an otherwise lovely day, and the fear of having nothing took hold deep within each of us. Where would we go? How would we live? Having nothing seemed as inconceivable as not existing, and the two became inextricably linked in my consciousness. This constant sense of scarcity—a fear of impoverishment that had been part of our DNA before it became a self-fulfilling prophecy—was a feeling against which my father rebelled with cocktails and ever more aggressive spending sprees.

One afternoon we came home from swimming at the club to the surprising sight of my father sitting behind the wheel of a new Cadillac Seville parked in the driveway. Standing there with one hand on the gleaming silver hood, he waved us out of my mother's station wagon. He opened the doors and had us smell the rich red leather seats and try the fancy electric windows and the air-conditioning. Three-year-old Whitney, I remember, smudged the glass, and my father barked at him to go inside the house and wash his hands. I felt a sharp pang of guilt as he slunk off.

"Come over here, Franny," my father said. "I want to show you something." He popped an eight-track tape in a slot in the dash and suddenly Benny Goodman's jazz filled the plush interior of the car. My heart swelled as I inhaled the new-car smell. I was flying high on abundance, and on my father's glorious mood—all the more so because, of course, I knew it wouldn't last.

My mother frowned and went inside.

Ollie peered out from behind a curtain. She did her best to protect all of us from my father. Often, after crying and begging her on a Friday, Whitney would ride with Ollie on the bus to spend the weekend in her house in Detroit, at the intersection of Seven Mile and Livernois, near to where the riots had erupted and where the sound of gunfire outside still blended unremarkably with the general hum of traffic. Even as a toddler, Whitney was a survivor.

And certainly being cooped up with my father for an entire weekend, my mother off doing volunteer work or playing tennis, promised a roller-coaster ride of unpredictable and treacherous mood swings more threatening than the odd gunshot nearby.

But now, enveloped in the luxury of my father's new Cadillac, I was happy. My father turned the key in the ignition and began to back out of the driveway just as Whitney was coming out of the house, his little hands dripping wet.

"Want to go for a ride?" asked my father as he turned onto our street—Grayton Road—and began to drive away.

I looked back and saw Whitney's expression shift from bright expectation to confusion. Clearly, he would not be joining us. My mother pulled him back into the house.

My father's love of cars, photography, and collecting was matched only by his love of movies. Horror films were his favorite, and he took me to see them all, including *Night of the Living Dead* within a few years of its release in 1968. He would routinely screen 35 mm films in our living room, too, inviting guests for Sunday afternoon movie binges that often included the Italian horror classic *Suspiria*, preceded by, say, Laurel and Hardy shorts. We kids would crunch our butter-drenched popcorn, thoroughly absorbed in this or that bloody scene, while the adults sipped cocktails, the sweet smell of gin our frail link to the relative safety of the offscreen world. And all the while my father would be standing by the projector, glowing with happiness, as *Halloween* or *Dawn of the Dead* unfurled on the screen.

I both loved and loathed horror films. The suspenseful music, the false sense of happenstance, the way two girls would get separated in the forest, guaranteeing their graphic slaughter—it all left me feeling by turns helpless and elated, danger becoming fused with excitement in my young nervous system.

MARY KATHERINE ROBERTSON AND GAIL ROBERTSON, CIRCA 1939

(by Norman Robertson)

My mother loved to drive cross-country. She took us everywhere by car—Florida, Martha's Vineyard, New Jersey—running up the miles on the odometer even during the energy crisis in the seventies. We would stop to nap in rest areas along the highways, the police sometimes knocking on our windows to wave us on. If the trip required an overnight stay and no cheap motels were available, we'd sleep on a community center floor or in the backseat. If my father came along on the trip, we'd stay at a Howard Johnson's or a Holiday Inn—the lap of luxury—until the car finally rolled into our resort or rental house.

"Why did you *drive*?" my cousins Pierre and Freddy would demand when we arrived at the ocean-side resort on Sanibel Island, candy wrappers and Coke cans littering the floor of our car.

"Flying and then renting a car is a waste of money," my mother would tell them.

"It's more fun to drive, anyway," I would lie. "Besides, we got to go to Disney World."

Running out of gas was my mother's specialty. We'd sputter to a stop at the side of the interstate, and she'd take our hands and march us along the shoulder of the freeway to the gas station at the next exit.

A free-spirited eccentric trapped in the life of a 1950s housewife, my mother would have been a hippie had she come of age in the sixties. She spent her childhood in Llewellyn Park, a spacious, wooded residential enclave in central New Jersey, outrunning her oppressive nanny at every turn. Red haired, freckled, and perpetually Band-Aided on both knees, my mother was the tomboy in the family, while her much older sister, Frances, received the lion's share of their parents' attention. With her raven black hair and classical features, and her admirable skill on horseback, Frances was the image of the perfect wartime debutante, and was even pictured as such in a 1944 issue of *Life* magazine.

My mother's father, Norman Robertson, was an engineer of Scottish descent who ran a family business that produced hydraulic pumps for clients like Thomas Edison, the family's neighbor up the road. An amateur jazz pianist and the life of every party, my grandfather entertained his many friends with Cole Porter while they fed him cocktails at the piano. My grandmother, Mary Robertson, also loved parties, and she and Norman made an especially compatible pair. At the end of the night she would help him home; as a girl, my mother would find her father in the kitchen the following morning, puffy eyed and bathrobed, hunched over a pint of coffee ice cream. "He craved the sugar," she told us years later. "But the cholesterol may be the reason he died so young."

Even during the Great Depression, the family was well-to-do, although my mother never forgot the sight of homeless people lining up at their kitchen door begging for food from the cook. The image made a deep impression, fostering her lifelong devotion to frugality. When her father died in her nineteenth year, in 1952, my mother kept her inheritance invested in the stock market, never spending a penny, except to pay her college tuition. She enjoyed the trappings of wealth but, out of a deep-seated fear of ending up poor, chose to live modestly, often buying her evening dresses secondhand, driving inexpensive cars, and carrying purses until they literally fell apart.

My mother graduated from Sarah Lawrence College in 1956 and met my father just one year later, in September, at my aunt Bettina Stroh's engagement party. She'd driven her red Triumph convertible over from Chicago, where her sister Frances lived, to Grosse Pointe on a windy day with the top down, a yellow silk scarf trailing behind her, her rebellious red curls cropped short.

"That your little red car parked outside?" my father asked when they were introduced.

She told him it was.

"I've got the Mark VI saloon parked near your car."

"Is that so?" said my mother with a flicker in her striking green eyes. Unable to stand bragging, she must have felt compelled to put the charming but arrogant Eric Stroh in his place. "You'd probably love the Rolls-Royce I've got back home in New Jersey."

Caught off guard by this entirely, my father turned and walked away.

During the engagement-party weekend my mother went out with my uncle Peter, but when she returned for the wedding in December, my father was assigned to drive her to the rehearsal dinner, and by the end of the weekend, she found she preferred my father. "I liked that he was artistic," she said. "And that we both had a love of photography."

They were married the following June.

Ollie sat brushing my hair after my bath. "You all are so *lucky*, Frances," she declared suddenly, surprising me.

"Why, Ollie?" I asked, expecting something really, really big, like Christmas coming early that year.

"*Why*? Because, you's rich peoples. Thas why."

Rich? I imagined Rolls-Royces and white marble mansions, like the Beverly Hillbillies. People on TV were rich; my friends whose last names I saw on the backs of cars, they were rich, but we weren't. Nobody else had ever told me we were, so how rich could we be? Besides, my father had said he couldn't pay the ransom.

Moments later, I confronted my mother. "Mom, are we *rich*?"

My mother's profusion of freckles seemed to darken, her broad, beautiful face clouding over with something akin to anger, as if I had spoken a four-letter word. "We aren't rich and we aren't poor," she said with firm conviction. "We're in the middle."

This was reassuring to me. So we were normal after all, which felt safe; kidnappers only took children from rich parents, not normal ones.

If only I could make Ollie understand. I traveled through the house following the sound of the vacuum cleaner and found her in a bedroom straightening a bedspread. "Mom says we aren't rich and we aren't poor," I told her. "We're normal."

Ollie gave me a searching look, then laughed and shook her head. She smoothed the bedspread until every crease was gone.

Later that day, after she'd taken the chicken out of the oven and changed out of her crisp, white uniform into what she called "street clothes," I watched Ollie walk down our driveway and up Grayton Road toward the bus stop. What she had said to me—that we were lucky because we were rich—was something that she would never have said to my parents, and I wondered if she saw my parents the way I saw the Beverly Hillbillies, not as real people at all.

The traffic on Grosse Pointe Boulevard was light on a Saturday. I did not feel the need to turn my head each time I heard a car approaching from behind, straining my neck, to check that my outstretched legs were safe.

"Make sure to hold your legs out," my mother had warned when I'd climbed onto the book rack behind her bike seat. "Don't get them caught in the spokes."

We rode this way often in the summer afternoons, careening the grid of streets, my mother stopping to greet friends she spotted on the sidewalks.

We were headed to the library, where my mother would check out her monthly supply of books. "Ridiculous waste of money, buying books," she used to say. I loved the library's musty smell. Paging through magazines, I would be lulled into a trance by the rustling of newspaper pages around me, until my mother finally came to collect me. Down the library steps we'd walk in the afternoon heat to where my mother's leather-seated old bicycle leaned against the brick of a wall, unlocked. She bungee-corded the books to the book rack before I got on.

"You add almost no weight," my mother laughed as she began to pedal. "I'd hardly know you were there."

Through the tree-lined streets we rode, my mother's freckled thighs slowly pumping our wire-spoked wheels along the pavement, until we arrived home with her armful of books. Ollie would be preparing lunch when we came into the house, chicken livers and green beans or hamburgers spattering grease from the open frying pan.

When I became too heavy to ride on the book rack, my mother bought me a Schwinn three-speed with a red, white, and blue bicentennial banana seat, my most prized possession. Now I could ride anywhere in Grosse Pointe anytime I wanted, so long as I came home for dinner. My friends and I would tear around the neighborhood, barely observing stop signs, or hang around Schettler's drugstore, eyeing the Revlon products. When we became hungry, we'd head up to one of the clubs for a meal, signing the chits with our parents' names. My mother hated getting big bills from the club at the end of the month, so I always felt relieved if my grilled cheese ended up on someone else's check.

With my new bike I graduated to independence in the blink of an eye, but I would always miss holding on to my mother's waist, inhaling the gentle perspiration through her starched cotton blouse.

In the summertime, my mother spent her afternoons playing backgammon with her friends on the upper deck at the "little" club. Off limits to children under the age of eighteen and located just above the snack bar, the upper deck was a reliable retreat from the demands of parenting while offering a panoramic view of the club's pool, where Whitney and I spent our afternoons swimming under the idle gaze of the lifeguard or causing mischief in the locker rooms. (My mother subscribed to Dr. Spock's parenting theories, which included permissiveness and a "trust yourself" approach to the rearing of young children.)

When my grandmother came for her summer visit from New Jersey, she'd sit regally on a chaise by the pool wearing a navy-blue sleeveless dress, clip-on gold earrings, and a banded straw hat.

"Mrs. Robertson *looks* rich," Ollie sometimes said of my grandmother. "She and your daddy's mother, Mrs. Stroh. The two of them got the skin of rich folks."

My grandmother would turn the pages of her Somerset Maugham novel while Whitney and I played Marco Polo or did flips off the diving board with the other kids whose mothers reclined on towel-draped chaises, Johnson's baby oil slicked over their legs and arms. I would be tormented by the charcoal scent of burgers grilling in the snack bar, or the sight of Dusty Miller sundaes carried out to the pool in tall waxed paper cups—perks of club life that Whitney and I rarely delighted in.

My grandmother always chided my mother in the car after we'd left the club. "Those lifeguards are fast asleep behind their sunglasses," she'd say. "Someone ought to be minding the children by the pool."

"You worry too much, Mother," my mother would reply good-humoredly. "The children are just fine."

I loved when my grandmother worried about us. "The children look pale," she would say as soon as she'd arrived from New Jersey, a great wave of abundance sweeping through our house in her wake. Baskets of raspberries and lushly arranged grapes filled the countertops. Lemon cakes and exotic flavored ice creams came out after dinner. (So unlike my mother, who bought only apples, green bananas, and vanilla ice cream at the A&P, avoiding the more expensive shops.) My grandmother outfitted us for the season, too—stiff new shoes for Whitney, a new Lilly Pulitzer sundress for me—and replaced the astringent Dial soaps in the showers with her own creamy Dove brand, and we all laughed out loud in the evenings when my father shouted from the shower, "Gail, where's my *goddamn* soap?"

. . . .

My mother complained that my father's compulsive collecting absorbed most of his earnings and Stroh Brewery Company dividends. "Your father's spending is like a disease," she would say. But her efforts to curb his spending only made him resentful.

I imagined something eating away at him from the inside, like a tapeworm or a tumor. I wondered if it was contagious, if we'd all caught it; perhaps it was only a matter of time before the scabs would appear, the lost limbs, or atrophied muscles. Soon would we all be in wheelchairs like Uncle Dan, who'd lost all his muscle control like that famous baseball player Lou Gehrig?

Only this disease seemed to travel a mysterious route; bypassing our tissues, it simply hijacked our feelings, our perceptions. A constant sense of anxiety quelled only by the distraction of intense excitement. What was worse, the condition seemed as incurable as my mother's badgering was relentless, and sometimes I wondered if all that spending might, in fact, be my father's attempt at remedy rather than the disease itself.

"Always save for a rainy day" was my mother's oft-expressed motto. She sometimes called our private school to get extensions on the tuition deadlines, or notified the phone company that the payment would arrive late. My mother arranged for us to have the hand-me-downs from her friends' children and, during the years when my father gave her no money for our vacations because he had spent everything on his collecting, we drove the five hours to my aunt's house in Harbor Springs.

Though both my parents' families were well off financially, the Robertsons' handling of money was far more conservative than the Strohs', as were their values. My mother had drawn from her inheritance to pay for college; my father had drawn from his to procure a fleet of Jaguars. From a distance, his largesse may have appealed to my mother, initially, but up close the two of them were like runaway trains passing in the night, my father's reckless spending accelerating with time, even as my mother's fears escalated proportionately.

After swimming at the club, while all the other children signed chits for hamburgers and Cobb salads, my mother usually took us home for peanut-butter-and-jelly sandwiches and brought us back afterward.

"Why can't you eat *here*?" our friends would ask and, feeling shamed, I lied and told them that we preferred the lunches our mother made.

I believed my friends at the club were more worthy than I was, and certainly luckier. They ordered club sandwiches and Cokes with a sense of abandon that I envied, while, on the occasions when I was permitted to order lunch, I enjoyed my hotdog and French fries—or my scoop of Stroh's peppermint ice cream—with a guilty pleasure that bordered on the illicit.

If I questioned my mother's policy, she reminded me, "The Carmichaels had to resign from the club, you know, because those kids ran up a bill so long, it arrived at their front door in a shoe box!"

My mother's anxieties fueled my own, and as a preteen I shoplifted clothing and makeup from the local department stores. I even stole my first trainer bra. Part necessity, part sport, my thievery had begun in first grade, when I'd occasionally taken pencils and erasers from my classmates' desks, and later at sleepaway camp, where I'd stolen a pair of Dr. Scholl's from a neighboring cabin, only to return them later under a cloud of guilt.

My middle school friends and I would take turns distracting the sales girls while one of us slipped the desired items into a canvas tote bag, our hearts revved up on adrenaline. Somehow my mother never thought to question my burgeoning wardrobe, even after I was caught and she'd had to retrieve me from security at Hudson's Department Store. Perhaps she was gratified, albeit unconsciously, by this demonstration of self-reliance.

At those moments when her worry overwhelmed her, my mother collected clothes from the street. As with backgammon, her compulsion became a kind of hobby. I'd come home from school to find piles of found clothes, laundered and folded or ensconced in dry-cleaning bags, stacked in neat piles on my bed.

"Can you believe I found this on Cloverly Road?" my mother said, bursting with excitement, as she held up a tired, old blue cloth coat. "Someone's moving, and they threw it out by the curb. Have you tried it on yet?"

I picked up the coat. "I don't like the style," I said. "There's too much padding in the shoulders." I'd grown accustomed to turning down such gifts, always taking care not to hurt her feelings.

Undaunted, my mother would deliver the clothes to the Thrift Shop, the Grosse Pointe hub for secondhand finds, where she volunteered on Saturdays. Then she'd bring home a bag of secondhand clothing for which she had exchanged the original garments, and I'd have to come up with a whole new set of excuses for why I didn't want to wear them.

"I have enough clothes," I'd finally tell her, gesturing toward my closets.

That usually ended the conversation. My mother foraged, I stole. Each of us had figured out our own way of coping with my father's disease.

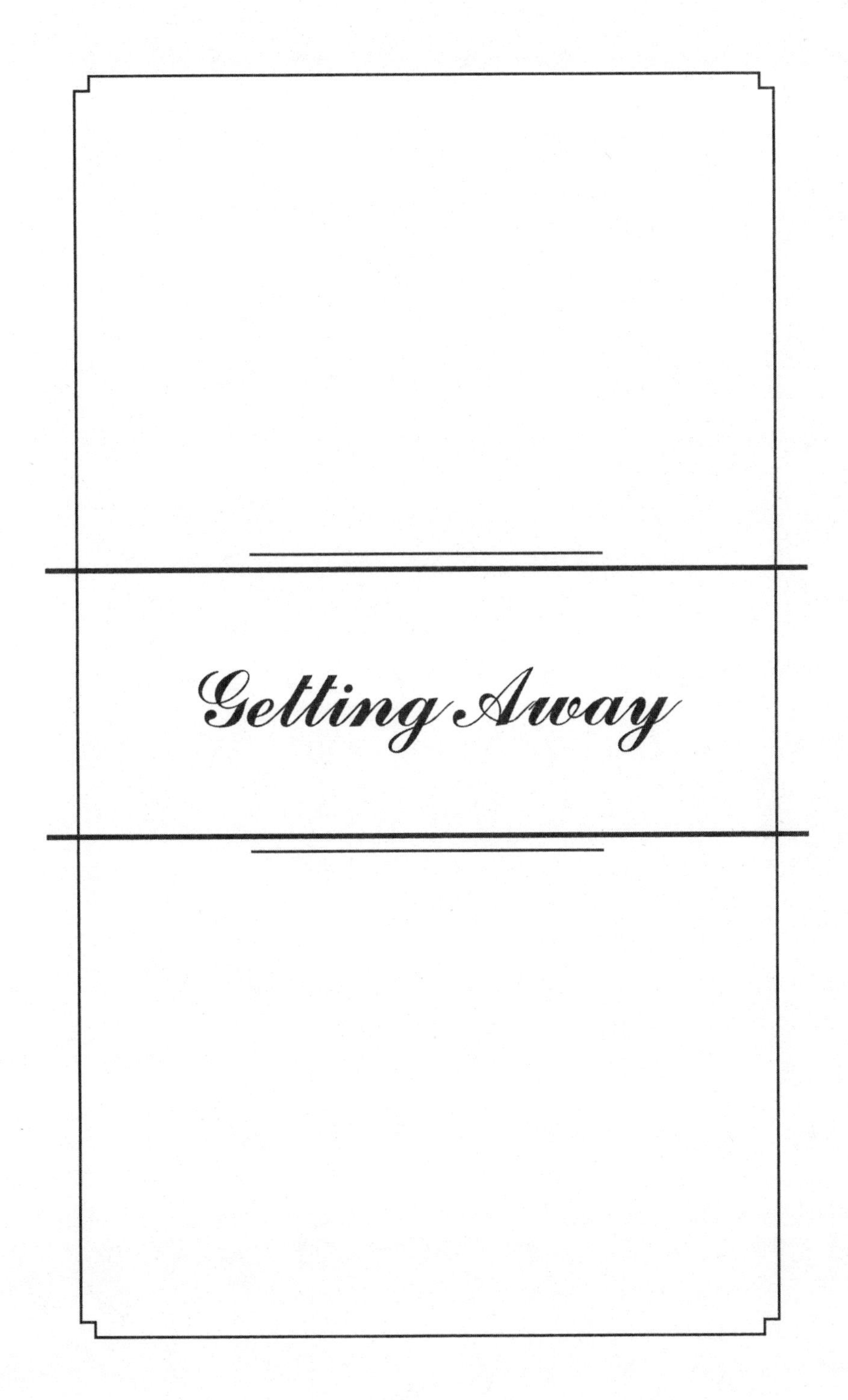

Getting Away

CHARLIE STROH, 1981

(by Eric Stroh)

Grosse Pointe, 1980

The December light faded so suddenly I could hardly read my own words. Rather than switch on the chandelier, I slid my high school application essay across the dining table closer to the bay window. Snow was beginning to fall. The empty house creaked around me as I bore down on my paragraphs, determined to get down exactly how things had felt the summer before, when everything changed, it seemed, overnight.

I wrote about my parents' faces—pale and swollen with sleeplessness—and the knotted feeling inside my stomach. Something terrible was happening: my mother had given up playing backgammon; my father had stopped leaving for work. I described the hushed voices, the closed doors, my gnawing sense that everything would come apart at any moment, that only a barely discernible tensing of all my muscles might hold it together.

My parents sealed themselves in the library for days.

"Whatever you do," my father said as he pulled the door behind him, "do not come in here."

Whitney and I sat on the porch watching TV, our blank faces masking our alarm, buoyed at least partly by *The Brady Bunch*, *Bewitched*, *Happy Days*. My younger brother's auburn hair was oddly disheveled, his trousers an inch too short. How I envied my older brothers, both of them off at college, Charlie a sophomore and Bobby a senior.

On day three my parents emerged: drained, older, yet united in their conviction that we should know the truth.

"It's so awful to have to tell you this," my mother began in a cracked voice, the puffed wedges underneath her eyes by now a deep purple. "But it's important you know: your brother Charlie is a drug dealer." Her eyes filled up with tears and she looked away.

My father dragged on his cigarette dismissively. "We're taking him out of college. Putting him into the Marines to clean him up."

As my mother wept my father put his cigarette into the ashtray and gently rested both his hands on her shoulders. I couldn't remember the last time I'd seen them touch.

"You must never mention a word about this to anyone outside the family," my mother said to Whitney and me with unusual sternness, wiping her cheeks with the back of her hand. "Nobody at all."

I felt the news and accompanying emotions seal themselves off inside my body with the ease of a closing elevator door. Drug dealers were something you saw on TV, not in my own family. I remembered an episode of *Starsky and Hutch* where the drug dealer lived in an abandoned apartment on the outskirts of town. Starsky kicked in the door while Hutch aimed the gun.

I turned on the chandelier so that I could reread my essay. Outside the snow was falling harder now, and a few stray sparrows pecked aimlessly at the frozen ground.

My last winter in Michigan.

Next year I'd be gone—away at boarding school for ninth grade, and away from this house. I'd been waiting to go since sixth grade, counting down the years impatiently.

The applications all asked for an essay on an experience that had changed my life. And so while other eighth graders wrote about their golden retrievers dying, I wrote about Charlie's drug bust and what it had done to our family, the shame and silence spreading from my parents to us, and then into just about every aspect of our lives.

As I wrote the story, I felt stronger, clearer—separate from those events for the first time. I wrote about the tension in the house, breaking it down into scenes with characters and dialogue, constructing not what actually had happened, but something that felt even realer than that. I wrote everything I'd been forbidden to say, everything that gave me back my voice. I wrote draft after draft, trying to get at the truth.

Charlie had been selling cocaine, a drug I knew about from *Time* magazine. I'd seen pictures of it, lines of white powder on the cover. My parents had heard the news from Charlie's college dean earlier that year, in the spring. He was expelled, and, under pressure from my parents, immediately enrolled in Marine boot camp in San Diego, leaving in early

June. But as June passed into July, everything kept changing. And the tension in the house got only worse.

Coming home from day camp or tennis lessons, Whitney and I watched TV in the library, or rode skateboards up the street with my cousins Pierre and Freddy. In his universal attempt to avoid my father, Whitney routinely asked Freddy if he could spend the night at his house, but Aunt Nicole sometimes locked the kids outside while the house was being cleaned, or while she napped, and for hours at a time no one would know where Whitney or Freddy had gone. They were nine and ten at the time.

When Whitney finally came home, he'd steer clear of my father even as he sought approval by doing small chores around the house—unloading the dishwasher, say, or feeding the cats. But my father would simply complain about the direction the forks faced in the cutlery drawer, or the ratio of dry cat food to canned food in the cats' bowls, and Whitney began to wear a permanent expression of defeat.

I usually stayed in my room playing Led Zeppelin and Rolling Stones albums over and over. Sometimes my father would throw open my door and shout, "Turn down that goddamn rock 'n' roll!" and I'd grab the volume knob so fast the record would skip. Later, I'd turn it all the way up again, partly to block out the eerie silence in our house, but also eager, in a way, for any interaction with my father.

I often found myself thinking about Charlie, who had turned strange over the last couple of years. Once tanned and vital, his face had grown pale and blemished, his eyes flattened like old decals. When he was home, he was, more often than not, on the phone or outside the house talking with people I didn't recognize, people who came up our driveway in their unfamiliar cars while my father was downtown at work. Charlie referred to the visitors as his "good buddies," but he never had much else to say about them. He crept around the house, always appearing busy, any closeness I'd felt with him utterly replaced by a disquieting distance.

Charlie wasn't the only big change under way. The Stroh Brewery was acquiring the F&M Schaefer Brewing Company of Pennsylvania. My father rarely talked about what was happening at work anymore and never wanted my mother to throw parties for his brewery colleagues at our house. He stopped flying to Hollywood, and soon new Stroh's Beer commercials were on TV—commercials that seemed to surprise my father as much as the rest of us. In one, the outlaw Jesse James held up a stagecoach for a case of Stroh's. It was exactly my father's kind of ad, but he walked out of the library after it aired, without saying a word.

In the evenings after work, my father often stayed out at the bars, or ordered pizza to eat in front of the TV while the rest of us had dinner in the dining room. On the rare occasions when he spoke to us at all, it was to shout.

My mother carried on as if everything were still normal, taking us for a swim at the club while she played her backgammon, out for Chinese food once a week, on errands to the dry cleaner, the grocery store, and the bank.

"Do you know what Charlie does in the Marines?" my mother said one evening in August as she placed plates of steak and potatoes in front of Whitney and me. "He wakes up at four a.m. and does a hundred push-ups. Then he cleans all the bathrooms on the base before breakfast." She forced a smile. "Don't worry, Charlie's going to be just fine."

"Does Charlie *like* being a marine?" I asked.

My mother gave me a puzzled look, as if liking the Marines were completely beside the point. "He's there to clean up his life," she said finally. "Put himself back together."

My mother's optimism was contagious, and we all believed in the quick fix the Marines would provide for Charlie. I pictured him picking up the pieces of his life, like so many shards of broken glass, going back to college, and eventually working at the brewery—the path expected of every young man in the Stroh family.

At the end of the summer, my parents flew out to San Diego for Charlie's boot camp graduation and stayed with friends in La Jolla.

My father was smiling as he loaded their suitcases into the car. "Charlie's a tough guy," he said proudly. "It's not everyone who can get through boot camp."

On the day of the ceremony, as I heard later, they waited in the auditorium, holding their programs, eager to see Charlie graduate into his new life, desperate to put the whole ordeal behind them. But when the Marines finally filed in, clean and spiffed up in their blue uniforms and broad-brimmed Marine hats, my parents noticed that Charlie was not among them. Confused, my father walked up the aisle to look into the hallway. There was Charlie, stiff with fear in his uniform, handcuffed and surrounded by several federal agents.

My father went back into the auditorium and took my

mother out by the arm. Charlie was gone. Done for, they drove back to La Jolla, my mother in tears, my father shocked and humiliated, both of them determined to conceal from their hosts what had happened. Their worst nightmare was unfolding: Charlie's drug dealing would surely make it into the papers now, especially if he ended up in prison. Everyone in Grosse Pointe would know—everyone *everywhere.*

My parents hired a Marine criminal attorney and brought Charlie home on bail. He'd been arrested, it turned out, for crimes he'd committed before entering the military; in the Marines he'd managed to stay clean. His officer wrote a glowing letter to the judge about Charlie's achievements in boot camp and about his changed life.

In the court hearing, taped recordings of our family's phone conversations were played out loud in the courtroom. The Feds, it turned out, had bugged our phone line for eight months the previous year. My parents sat next to Charlie's attorney, listening to recordings of me gossiping about boys and parties with my middle school pals. Then came the recorded drug transactions—scores of them.

Ollie admitted to my parents that she had noticed Charlie handing off packages to cars that came up our driveway. My father made inquiries and discovered that the son of some people he knew in Grosse Pointe—another college coke dealer—had reduced his own sentence by tipping off the Feds to Charlie.

After a protracted trial, Charlie got off with a large fine and probation. No prison sentence. As for the media coverage that my parents had feared, it never materialized. The

Marines, it seemed, had done the trick. And Charlie still had his four-year tour of service ahead of him.

I sealed my last application inside the envelope along with my finished essay. Everyone else in the family had gone to the airport to pick up Charlie, even my father. My brother was on leave for the Christmas holiday from Camp Pendleton, where he'd been stationed after boot camp.

Outside, I walked in the dark down the winding, snow-covered driveway toward our mailbox, the four envelopes snug in my gloved hands—envelopes addressed to Choate, Groton, Taft, and Brooks. The sky was blacker than I'd ever seen it, and I crunched along the tire tracks in the snow until I came out into the street, where the mailbox stood. As I pulled it open, I imagined myself throwing open the gates of my life. A great wave of hope swelled within my chest. Soon my story would be out in the world. I imagined its debut as a loud, crashing sound, like the aftermath of a bomb. I wondered if anyone else would notice the explosion.

FRANCES STROH, 1982

(by Eric Stroh)

Carrying my trunk and an Oriental rug, my mother and I climbed the cement stairs of the dormitory to the second floor and found my room, a tiny cell with flimsy metal-framed bunk beds, two dressers, and two desks. Old lead-paned windows looked out onto the green rolling hills of western Connecticut, punctuated by clusters of oak trees and, beyond them, the school's perfectly groomed athletic fields.

We unrolled the rug across the gray cement floor, instantly brightening the room.

"Cool rug," said a narrow-faced girl wearing a man's felt hat. She'd come in and sat down on my trunk unannounced. "I'm Jen."

"Jennifer Victoria Fairchild?" I said, shaking her hand. I'd gotten the school letter with my roommate's name the week before.

"That's me," she giggled. She went over to the bottom bunk and unrolled a poster: Bob Marley smoking a cigar-size joint. "Where should we hang *this*?" she asked.

My mother shook her head with resignation, pushed my trunk into a corner, and gathered her purse.

I laughed, feeling a spike of excitement. That poster represented a whole uncharted world. I had smoked pot only once, in eighth grade, but I'd never actually been high. I was determined to change that this year, in spite of my parents' warnings that pot led to "harder stuff." Now, of course, they could point to Charlie as an example, but I wasn't like Charlie; I was smarter than he was—I was sure of it—I could break the rules and still come out on top.

Soon my mother would get into the car and drive away and everything would become possible. My heart quivered at the thought of her leaving. She would worry, I knew, but only for an hour or two. I would miss her; I would miss a lot of things, and yet my whole life, I realized, had been preparing me for this moment. I saw my childhood spinning away from me like so many lackluster previews before the feature itself. I looked around at the blank walls of our cell. "Anywhere," I said.

"There are other cool posters we can buy in the Village," Jen said. "Taft charters a bus to New York for long weekends."

The campus of the Taft School, founded in 1890, was made up of old neo-Gothic buildings arranged on hundreds of acres of greenery. The school had only begun accepting girls in 1971, ten years before my arrival, around the time it eliminated coats and ties. The most flamboyant students now were throwbacks to the sixties, with their long hair, hippie beads, and Mexican ponchos, while the faculty seemed permanently frozen in the conservatism of the 1940s or '50s, the era they'd begun teaching there, in most cases. They seemed to have unfurled like ivy from the school's stone walls and spires, these tweedy, pipe-smoking teachers, and appeared entirely unprepared for the hopelessly stoned, guffawing students that passed them in the halls. The less passive among them cast their disapproving gaze on these bohemian kids whose politics were so vastly different from their own. Like flies in amber, the warring forces of straight America and the hippie movement appeared perfectly preserved at the Taft School.

There were only thirty-five boarders in the freshman class, and ten of us banded together in a tight-knit group whose closeness surpassed our own sibling bonds. Among the girls, we shared our clothes, our cigarettes, our boyfriends, our houses and apartments, and, soon enough, our alcohol and drugs. Almost overnight, in the absence of parents, we became one another's family.

Music was our religion. The Grateful Dead, the Doors, and Neil Young—they were gods. And the Talking Heads, the Police, and the Clash, demigods. We all went out for sports, pulled all-nighters to study for exams, and never missed a class. But in our free time, we asserted our newfound independence by violating each of the school's rules, one by one.

The doors to the dorm rooms were kept unlocked, except when parties were under way. The junior and senior boys regularly supplied us with pot and shrooms. Day students provided alcohol from their parents' liquor cabinets or purchased on trips to Waterbury.

Taft had a two-offense policy: two drinking and/or drug

offenses and you were out. The fact so many "Tafties" were the progeny of prominent Wall Street financiers, famous authors, and brand-name families hardly mattered. Among us kids, the individual was judged by how much he could party—how well he played the game of risk—and still do well in the games of school and sports. Straight As, varsity teams, good taste in music, and a robust drug habit, that was what landed you on top, socially. Surviving a "bust" made you legendary.

The academics at Taft mattered as much as anything else, though. I spent my first exam week snorting crushed NoDoz while studying until dawn, my friends and I taking cigarette breaks every hour down in the butt room—a room-size ashtray in the basement of Mac House. My favorite course was third-year Latin with the preternaturally acute Mr. Cobb, whose famous line—"Whenever you have a fifty-fifty chance, ninety-five percent of the time you'll get it wrong"—seemed to apply to everything.

In the dorm we'd stay up late discussing Camus while drinking home-fermented cider (produced by adding yeast and three successive rounds of heating next to the radiators and cooling in the snow), or collaborate on algebra problems, the Ramones beating a manic rhythm in the background, while packing for our next weekend getaway. Because, with Taft's chartered buses and lenient travel policies, New York quickly became our off-site playground.

Nearly all my friends had either grown up in New York or had a parent with an apartment there. Sasha lived just off Washington Square Park on Waverly Place, Liv on Eighty-First and Riverside Drive, Cece on Park and Sixty-Sixth, and Feren's dad in a high rise on the East River.

We stayed at these apartments on the weekends, when my friends' parents were out at their country houses, and spent our days trolling the West Village. The Village dazzled with its array of head shops, art house cinemas, and the most coveted suppliers of bohemian wear in Manhattan—Canal Jean and Reminiscence. Indian print T-shirts soon replaced my Alligators; dangly silver filigree earrings replaced my tiny gold studs.

We shopped with the cash dispensed by Taft's accounting office, our expenses later billed to our parents. Despite my mother's pessimism, the Stroh Brewery had expanded nationally by acquiring the much larger Schlitz Brewing Company—described by the *Wall Street Journal* as a "minnow swallowing a whale"—and money came my way more easily now; I could see our beer brands in all the liquor stores in New York, the empties in Central Park and even on the Taft campus. Still, handing over those crisp ten- and twenty-dollar bills in West Village shops, I got a rapid-fire charge of adrenaline guilt, as if spending money were as risky as shoplifting or smoking pot. I'd witnessed my father's freedom with money, of course, but never without my mother's sportscaster's commentary. I'd internalized both somehow, so that spending became just one more naughty thing I could do.

Fifteen-year-olds with pockets full of cash, we felt as if we owned the damn city. We dressed up and went out to Upper East Side bars, practicing our classroom French on the bartenders, ordering martinis. We took cabs downtown, had dinner at John's Pizza in the Village, and danced at Studio 54 or Danceteria, flashing the bouncers our fake college IDs, procured at Playland on Forty-Second Street. We bought pot in Washington Square Park to take back to whichever apartment we happened to be crashing in, and if no one was selling the good stuff, we called Choo-Choo.

Choo-Choo brought the drugs to your building, uptown or downtown, like a train carrying precious cargo. A scrawny old punk rocker, he'd show up wearing peg-legged black jeans, a wifebeater, and a spiky dog collar–style belt down low on his waist.

One Sunday after a particularly late night, a group of us sat around Liv's penthouse on Riverside Drive waiting for Choo-Choo's delivery. When the doorman telephoned up to announce, "Miss Goodman, Mr. Choo-Choo is here," Liv and I went down in the wood-paneled elevator with our wad of collected bills.

Choo-Choo was waiting next to the leather sofa in the lobby, I remember, sweating behind his sunglasses. The doorman politely looked away when we pulled out the cash.

Upstairs, we sat in the living room drinking really strong gin and tonics and smoking Marlboros, the baggie of green buds on the coffee table in a nice crystal ashtray, right next to the *Times*—with its front-page article on Choate Rosemary Hall. Two Choate students had been arrested at Kennedy Airport for trafficking cocaine up from Colombia, and fourteen others had pleaded guilty for financing the trip. The article seemed to encapsulate a multitude of truths about our kind of crowd; we took risks—big ones—as a diversion from our boredom. It wasn't about running away "Summer of Love" style; we worked from within the system, using our own privilege as a launching pad. The trick was recognizing where the lines were and then letting someone else cross them. One of the traffickers, see, had been the son of a truck driver, hell-bent on fitting in with the privileged set at Choate, and . . . he'd been used. Just as Charlie had been used. They were the line crossers.

"Can you *imagine*?" said Feren. "You'd have to be, like, *so* fuckin' ballsy to go for that."

I sipped my drink and studied the faces of the kids involved—kids just like us. I couldn't believe they'd done it. "I almost *went* to Choate," I said. "I wonder if they'll go to prison."

"That so easily could have been you, Franny," said Jen, absently dropping her ash on the rug. "You know? I could totally see you doing it."

I thought of Charlie—now stationed in Okinawa—and I felt an inviolable boundary within myself. "Nope, I'd never have gone for that." I already understood, at age sixteen, the elusiveness of the line between a life of privilege and a life in prison. Soon after Charlie's trial, Michigan had raised the penalty for dealing cocaine to life without parole.

"Would you?" asked Sasha, looking over at Feren.

"Me?" asked Feren jestingly, laughing her semi-insane laugh. "Not a chance."

But I wasn't convinced. Feren was the wildest of us all. Over spring break she'd apparently been cavorting with some French sailors who'd docked in Nevis, the West Indian island where her mother lived.

"I wonder what will happen to them," I said, glancing down at the photo. "They're so screwed; they'll never get into college."

"They'll end up in reform school," said Liv in that endearing deadpan monotone of hers. "As far away from drugs as you can get."

Feren got up to replenish her drink. "I don't know, Liv. They probably have great drugs in reform school. Like they do in prison."

"Better than at Taft?" I asked. I couldn't imagine a place where drugs were more available or more intrinsic to the culture. In Grosse Pointe some kids I knew were starting to use coke, but at Taft you'd have to hide under your bed to avoid being implicated in drug use, and you'd probably find someone's stash while you were under there.

The irony was, my parents had sent me away at least in part to protect me from drugs. As it turned out, getting high was just such a key part of life at Taft—an essential step toward becoming an adult, an instant form of self-reinvention, and certainly a step away from a childhood that was best left behind. I talked with my parents once a week on the pay phone in the hall, but with all the kids gone from the house except Whitney, I imagined the scene at home was rough. My mother's persistent cheerfulness in the face of my father's emotional decline was enough to keep me in New York for half of all my vacations, either with friends or at my grandmother's house in New Jersey. I hoped Whitney would survive until he could go away to school, too.

Our "dorm mothers" were too detached to snoop—at least until sophomore year, when I roomed with Sasha. We'd chosen the room because no teachers resided on that hall, and the set of purple psychedelic curtains framing the window at the end of the hall, just outside our door, was a main attraction. We loved those curtains, with their absurdly bold swirling patterns, so retro 1960s, so symbolic of the ironies of Taft. Once Sasha took them down and donned them as a cloak she wore to sit-down dinner.

Pamela, a tall, skinny blonde from South Bend, Indiana, usually came down to do bongs with me. We used a "hit towel"—a regular white towel, dampened and rolled up for maximum absorption, into which we blew the smoke to avoid stinking up the room—and we sprayed Ozium in the air, as an added precaution.

But then Jan Coleruso, a newly hatched teacher from Yale, started knocking on our door during study hall, asking for aspirin and tampons, inhaling our room's aroma as she stood in the doorway, her running shorts sprouting thick, muscular legs.

"Why is she *stalking* us?" I complained to Sasha as soon as she'd left.

"Um, because you're a pothead?" Sasha would say wryly, grinning as she went back to her book. This was the routine.

Being practical, Pamela and I changed our schedule: we brushed our teeth, smeared Clearasil on our faces, and did bongs every night before bed, avoiding smoking during study hall hours.

Then one Saturday night my luck ran out. I'd been playing Quarters at an off-campus party, and my friend David, a day student, dropped me back on campus past the midnight curfew. The doors to the dorms had already been locked, and everyone's lights were out except Coleruso's. I knocked loudly and waited. I heard footsteps on the cement stairs, doors creaking open, and then my heartbeat pounding in my ears when I saw her through the porthole, turning the key in the lock, a gaseous cloud of beer and cigarettes wafting into the vestibule when I stepped in. I knew I smelled like a frat bar.

"Sorry I'm late," I said.

Coleruso just stood there staring at me. Twenty-two and new to Taft, she had no idea how to carry out a bust. Having shadowed me for months now, she was freezing.

I signed in on the roster and swiftly retreated to my room before she could figure out what to say. Three days later, she turned me in to the dean. I was suspended for two weeks. My parents took the news in stride. A little beer wasn't a big deal to them, although it was clear they hoped I would clean up my act after my suspension, which I spent in Grosse Pointe with Whitney and Ollie while my parents vacationed in the Bahamas.

Two weeks in Michigan was a long time, particularly in February. I couldn't call any of my friends in Grosse Pointe, because my parents didn't want anyone to know I'd been suspended. Confined to the house, I spent the mornings keeping up with my schoolwork and the afternoons watching HBO with Whitney.

I missed getting stoned with my friends at school and would make do smoking a bowl myself in my room at the end of the day. I'd sit in my window seat watching the flurries blowing around outside, the house dead silent, Ollie downstairs roasting a chicken the way she used to do for us. Whitney would be in his room doing homework. I wondered if I appeared to him the way Charlie had once appeared to me—pasty skinned and preoccupied. Whitney had been alone for a year and a half, and this was our chance to reconnect; instead, we were both holed up in our separate quarters. I'd given him a couple of Neil Young cassettes, and I could hear "After the Gold Rush" floating down the hall from his room.

I took another hit, holding the lighter in the bowl until it burned my thumb, and exhaled. I cracked the window. A few flurries of snow blew in, melting instantly.

True to form, Ollie never mentioned the daily cloud of pungent smoke to my parents when they called to check on us, or let on about much of anything. She'd always been on our side, throughout everything that had happened, and I loved her for that.

We stood under the bleachers at a lacrosse game—Taft versus Hotchkiss—while the crowd above us roared.

"My brother was busted by the Fed for dealing coke," I told my friend Trey. My head felt light with the beer we were sharing.

"No shit, really?" He took the last drink from the can, then tossed it on the ground. "The fuckin *Fed*?"

"But don't tell anyone." I tucked my hands into my jeans pockets. "My parents made me swear I'd never say a word about this."

"I swear," said Trey. "But that's, like, crazy."

We'd been hanging out lately, getting stoned in the woods, smoking cigarettes and kissing behind the science building after vespers and dinner. I'd hung out with him one night in the fields while he tripped hard on mushrooms, laughing aloud at the stars. He had a high school band—Space Antelope—that played twangy Grateful Dead–inspired tunes.

"You're the first person I've ever told out loud," I said. "I guess I needed to tell someone."

Trey looked at me and grinned. "Cool. So now you're lighter, right? Now you don't have to worry anymore—you're free." His auburn hair caught a splinter of sun, turning it gold. His lashes were blond, almost invisible.

I leaned against the rough edge of the bleachers. The crowd stood up and cheered, a riot of stomping feet and shouts.

"Now I'm free," I repeated.

I wondered if this could be true. I never felt free for very long, only for a few days or weeks before the heaviness came back. Whenever I had that feeling of lightness, I knew it wouldn't last. Which meant I had to do something, drum up some new excitement, to keep ahead of that terrible weight.

. . . .

For ten hours I pretended to be asleep in the back of the car with my trunk. My father drove silently, stopping only for gas and food. I knew he must be too angry or too heart-broken to speak, and his silence was, in fact, a relief to me.

Taft was throwing me out. I had only one documented offense, even counting the bottle of vodka found in my room that I swore I'd never seen before, but the school, reserving the right to change the rules, had decided I shouldn't come back for junior year; they saw me as a ringleader of sorts, my influence spreading to innocent freshmen like Pamela, and so I was being excised, like a cancer.

I remembered my mother telling me about when Bobby was expelled from Kent, how he and my father had cried together at the airport. Charlie's expulsion from South Kent had followed, then his college expulsion, and now this. Would *I* end up in the Marines?

The car came to another stop. I heard my father pumping gas. A few minutes later, he came back with what smelled like McDonald's burgers and fries. I heard him trying to hand a bag to me in the back but even now I didn't open my eyes.

Tears washed over my cheeks, perhaps at the kindness in my father's gesture, what seemed almost like forgiveness. I was losing everything—my friends, my room, my independence—but . . . at least my father was with me there, silently, on my side.

He had driven from Michigan to talk the school out of expelling me, but the meeting apparently had not gone well.

This was why I was "asleep" in the back of the car, and why I would have to spend the next two years back in Grosse Pointe until I could go away to college.

Ever since the bottle of vodka had been found, I had cleaned up, avoided all the spring parties out in the fields, the smoking sessions in the day-student locker room, but the headmaster had told my father, "Sorry—too little, too late."

I knew they were making an example of me. My friends whose fathers had attended Taft, they had all been given warnings, while I was being expelled. Charlie had been a scapegoat as well—getting clean only to suffer the repercussions of old crimes. The fact that life was intrinsically unfair lodged itself at the center of my chest, like a well-mortared brick. I loved Taft; I'd finally lived my life fearlessly, everything within my grasp. With no one to stop me, I'd ordered the proverbial club sandwich—and I'd devoured it whole. Now those old feelings of unworthiness were creeping in again, and I wondered if I'd even deserved any of the happiness I'd felt over the previous two years, the frenzied sense of freedom.

Granted, there probably wasn't a single handbook rule I hadn't violated, with the exception perhaps of plagiarism or cheating, but my grades had been good and I'd been a strong athlete—varsity ice hockey, varsity tennis. I'd seen plenty of kids get away with more than I had. In the boys' dormitories I'd seen bongs as tall as standing lamps, with lamp shades placed on top as their only disguise. I knew students who were regularly invited into a certain teacher's apartment for cocktails, and boys who got caught red-handed with drugs and faced no consequences. The omnipresence of drugs and booze at Taft had taken nearly all of us up in its mischievous embrace. The chartered buses into the city, the free-flowing cash, the "chaperoned" theater trips, the dinners at Beefsteak Charlie's with free pitchers of beer. We'd all partaken.

And then they'd changed the rules on me.

Hearing a crackle of plastic, I opened my eyes: my father, opening a pack of cigarettes. I sat up in the seat, and he eyeballed me in the rearview mirror, a slight smile on his face. He lit two cigarettes with the car lighter and handed one to me.

I took it silently. The last time he'd given me a cigarette I'd been ten, also riding in his car. I took a long drag on it and looked out the window at the blankness of Ohio, picturing my father and Bobby smoking at the airport, the tears drying on their cheeks.

We were almost home.

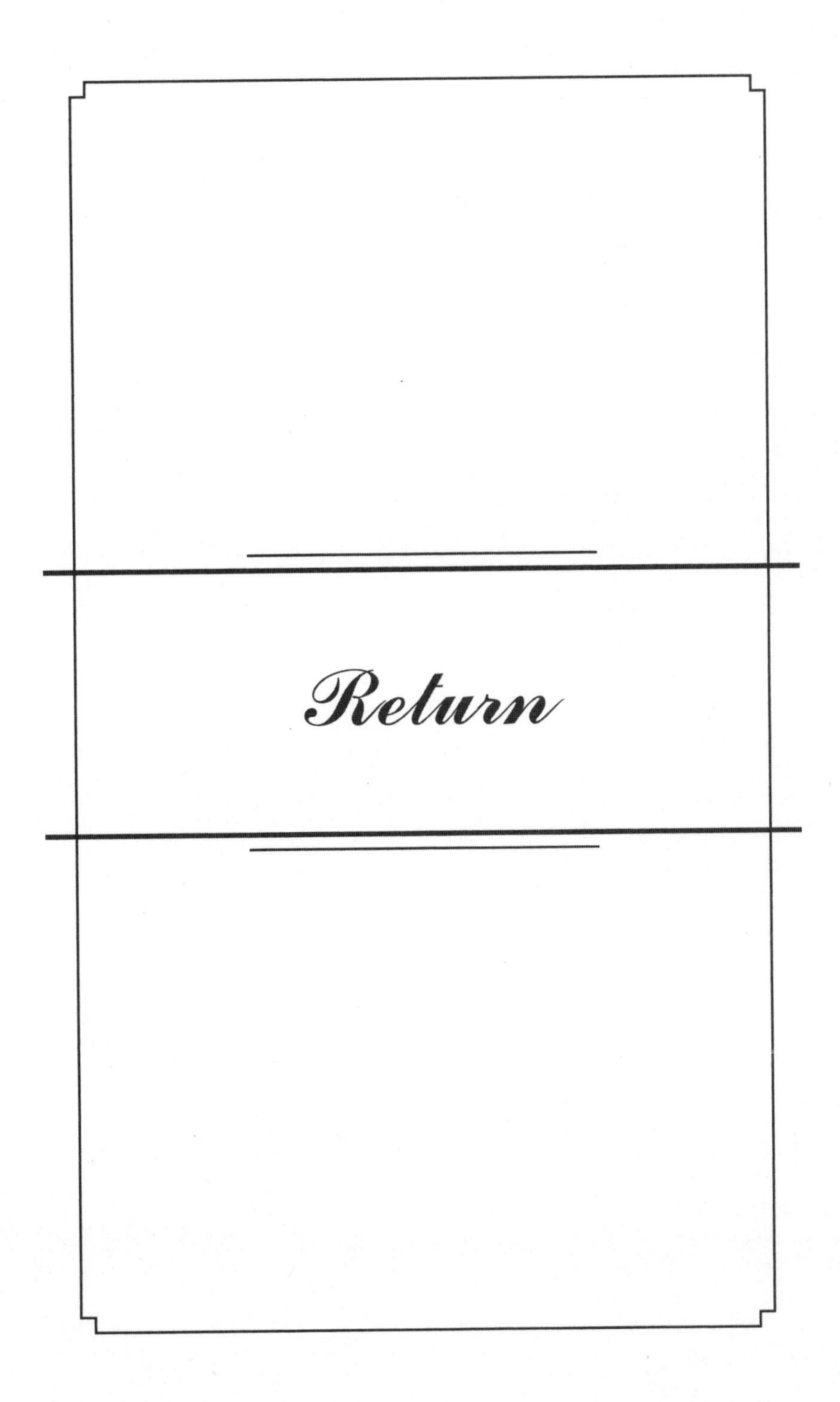

Return

WALL IN DETROIT, 1984

(by Frances Stroh)

Detroit, 1984

The potholes on Jefferson Avenue got worse every year. Five months of ice and snow and then the blasting April heat caused the pavement to buckle, then shatter. This, combined with the badly depleted coffers of local government, meant we had a piece of Swiss cheese for a road as we headed into the heart of Detroit to the abandoned Uniroyal Tire plant, an icon of the city's manufacturing past that was set to be demolished in just a few months.

We had been planning the excursion for weeks, speaking in muted voices in the hallways at school, lining up a car, discussing the pros and cons of various drugs. LSD, we'd finally agreed on.

Hobey drove while I sat in the passenger seat and Caitlin and Mike lounged in the back. Dilapidated buildings lined the street, and ambling jaywalkers crossed in front of our car. Hobey knew we were coming on, and he cranked up Bad Brains—hard-core raspy punk that made my skull ache. Grosse Pointe seemed a million miles behind us.

I observed Hobey's herringbone tweed overcoat, his trim buzz cut, his wicked smile—all so out of sync. Bags of provisions and gear were piled at my feet: cigarettes, beer, water, camera equipment. We passed a billboard displaying the tagline "Stroh's Is Spoken Here," and I felt a little ashamed. Stroh's was Detroit's beer, like Bud was for St. Louis, but it wasn't as if I'd done anything to deserve having my name up there like that. No one else noticed, or they didn't report on it if they did. My friends and I never talked about it, the wealth I'd grown up with.

Everyone in the car jammed to the music and smoked. We had Mickey's with us, not Stroh's. I opened one to calm my amped nerves. I was coming on fast to the acid. Outside the window, the streets visibly vibrated with the music while the car floated forward, a bubble of self-sufficiency, into the density of the forsaken city.

Toyota and Honda were winning the war. And Detroit, Detroit was an aftermath in the form of a city, a tragic defeat of all things American. Populations, industries, architecture—collapsing in on themselves like decimated ruins. The wreckage of this city etched itself into my genetic code, I could feel it, the patterns engulfing every shape and color.

I turned to Caitlin and Mike in the backseat. They smiled, their eyes melting pools of light. Caitlin's black hair blew sideways across her face. She had a fragile beauty, with the translucent skin, the shadow-encircled eyes, of a druggie. Mike wore his sheepskin vest, a silver medallion resting against his hairless chest. He put his hand on Caitlin's, holding on while his feet lifted off the floor, his red curls all liquid fire as he threw his head back, laughing at nothing at all, or maybe everything.

"Wow . . ." It was all I could think to say. On the street a homeless man shared his food with a dog. "Wow, I mean, look . . ." I trailed off, the waves of feeling so intense I knew they must have seen it, too—the open wound that was the world.

Hobey switched the cassette and turned up the volume on "Box of Rain," smiling over at me. Deejay as shaman. The gentle, twanging notes made the sun feel warmer, the sky bluer, the bombed-out city a kind of refuge.

Caitlin knew the way into the abandoned plant. An alley off the street led to an open side door that had been used by squatters. Old soiled clothes littered the filthy floor, along with broken glass, blankets, syringes. Entire walls had been painted black, as if to simulate night.

We walked in astonished silence toward a great wall of windows, the sun pouring through, a waterfall of light. Outside, the vastness of the defunct tire plant stretched before us like a forgotten city: a maze of streets lined with two-story buildings, railroad tracks, even a garbage dump where torched tires melted under the sun. Yellow-and-black signs reading Hazardous Waste had been posted on the sides of buildings. Concrete rubble piled up everywhere, as if an earthquake had toppled whole structures, forcing the city's inhabitants to flee.

We carried our cameras outside, into the maze. Caitlin and Mike held hands as we ducked under a barbed wire fence and walked along the train tracks. We pushed open an unlatched door, and our footsteps echoed inside a warehouse stacked to the ceiling with shipping containers. Dust hung in the air, catching the gauzy light.

No one spoke. I snapped a shot with my Nikon. "Day turns to dust," I heard myself say, breaking the silence.

The warning signs had been posted everywhere. Uniroyal was a forbidden zone, the site a toxic waste–dumping site long before the hazards were even well understood.

We walked deeper into the maze of buildings, taking more pictures. Pools of putrid water had formed from all the spring rains, some of them full of trash and weeds. We wove around them, separating and coming back together, laughing out loud when one of us splashed through recklessly. Mike splattered mud on Hobey's coat, and Hobey kicked a black leather boot into the water, soaking Mike.

"You asshole!" Mike shouted. He brushed the grime from his vest and arms. When the water had settled, a solid organic object—something dead—floated at his feet.

Hobey stopped laughing. "What the fuck."

Caitlin walked closer and poked it with a stick, turning it over. "It's a rat." Her voice sounded hollow. She held up her camera and snapped the shutter.

"*Je*sus Christ," said Mike, stepping out of the water.

I bent over to see it closer. The eyes were open and black, the body the size of a small cat. The fur rippled rhythmically, and I wondered if it might still be alive, but Caitlin's sweater was rippling, too. Everything with texture was alive.

I imagined thousands of rats all around us, rambling through sewer pipes or thrashing their tails inside little holes in the ground. I'd read a book once about the apocalypse that said only rats and cockroaches would survive a nuclear winter. They'd proliferate and take over the earth for millions of years until new species evolved. Even now, I thought I could feel things crawling on me. I scratched my arms up and down until I'd begun to produce welts.

A door slammed from somewhere inside the plant, and the sound ricocheted off the buildings in the silence. We all looked at one another. Had it been the wind? Anything could happen in here, and no one outside would know. People were murdered in Detroit every day.

I looked behind us and realized I wasn't sure how to navigate back out. "This is too weird," I said.

We lit cigarettes and listened to each other's labored breathing. Mike's curls looked artificially red under the now poisonous sun. He scanned the outbuildings of the walled city. "We should go," he said, his voice shaky.

Hobey put his hand on Mike's shoulder. "Relax, man. You're just having a bad trip, hunh?"

But I'd had enough, too. "Let's go back," I said. "We get caught in here, and we're screwed."

"No shit," said Caitlin, her pupils spheres of black velvet. She replaced the lens cap with a slender, jittery hand.

The magic had worn off, except for the bad kind. I thought I could feel the cockroaches under my clothes, could taste the radioactive particles on my tongue, like dirty pennies. And I suspected we weren't alone in the plant.

We took off, sprinting across the railroad yard, but I stopped suddenly and turned around, taking a last look.

The web of discarded streets and buildings went on forever. The rest of Detroit wasn't far from being like this—every windowpane shattered, every door hanging from its rusted hinges. Life had left this place. I felt my legs running toward the distant sound of feet hitting ground. I went around some warehouses and ducked under the barbed wire fence where I knew my friends had been, their dust still suspended in the air. I saw them waiting for me in front of the main building. We threw open the door and raced through to the street.

We stood panting next to the car.

"Let's have a beer," said Hobey.

My father was cutting the lawn when I arrived home, and both my parents' cars were parked in the driveway. I could see the flashing screen of the TV through the library window. Being around my parents while on acid was an experience I generally tried to avoid, but everyone had bailed early, saying they needed to get home for dinner. Only then had I even remembered it was Tuesday. We had skipped school.

I entered the house through the side door, went up the back stairs, and slipped into my brother's bathroom, where I knew no one would find me. The mirror caught my reflection; my eyes were all pupils, lashes grotesquely long. My skin looked orangey red. My geometrical bob was knotted from the wind, except where the hairspray had secured it. Hobey called that sprayed bob my Darth Vader helmet. I laughed out loud at the absurdity of this.

My skin was moving, pulsing uncontrollably, like the walls, the tiled floor, the grain of the beige carpet in the hallway. Atoms, constantly moving.

I sat on the toilet to steady myself, the wall cool against my neck as I leaned back. The smell of freshly cut grass floated up from the garden through the open window, the sound of the mower as familiar as my mother's voice. I was home, safe in Grosse Pointe, the wasteland of Detroit some infinite distance away. Our neighbors, the Fruehaufs, were readying their pool for the summer, while the high school tennis team at Liggett was vigorously hitting balls, their courts just on the other side of the fence from us.

I knew I would go downstairs soon. I would act normal, as if I'd been at Hobey's house since school had gotten out, listening to his band practice in the basement. I would plant myself in front of the TV, find a movie on HBO, and wait for my mother to announce that dinner was ready.

But then I heard another familiar sound: my father's voice. He was shouting over the lawnmower, probably at Whitney. Tuesday, I reminded myself again. My father would have been "at work," meaning he'd been drinking all afternoon downtown near his office, putting me in grave danger of having a bad trip if I saw him.

I went down the back stairs and out of the house, and climbed into the front seat of the Ford station wagon. As I turned the key in the ignition, I could still hear my father's infuriated voice coming from the flagstone path that led to the terrace and lawn.

. . . .

Out on the road I picked up speed. To the right was the country club golf course, a blinding streak of neon green. On the left were storybook estates—replicas of English manor houses, their sweeping lawns studded by towering trees. Twice as tall as the houses, their branches swayed like dancers' arms to an internal rhythm I couldn't hear, beckoning me. It was the waning acid talking, I knew, because the houses had invisible force fields keeping me at bay. In all their perfection, the homes looked as lifeless as the Uniroyal outbuildings, only the gardeners visible, in glimpses, to let you know any of it was even remotely real. Our house, a big brick Colonial we'd moved into when I was nine, was matchbox in comparison, if a good deal more inviting.

I passed the Williamses' house with the antique-car collection in their twelve-car garage and, with my one-handed steering wheel grip, stubbing out my cigarette in the ashtray, took the curve past Buhl Ford's house. It seemed every house on Provencal Road had a family just like mine: at least one fanatical, uptight parent, with a host of wayward, rebellious youth. Often enough one of these kids wrapped a sports car around a tree on the golf course, the tire marks on the putting green a haunting reminder in the weeks that followed.

All the kids on the street, except me, either went to Liggett—the private school right by our house, where I'd gone for grade school and junior high—or to boarding school. And since Liggett hadn't taken me back after I'd been expelled from Taft, I'd matriculated at Grosse Pointe South High School the previous summer.

"Public school," my mother had lamented, "Oh, Frances . . . *that* really is an embarrassment." I reminded her that Bobby and Charlie had also served time at South High after their boarding school expulsions. So I was in good company.

As I approached my cousins' house I could see both Uncle Peter's and Aunt Nicole's cars parked in the driveway, along with that of Gwendalyn, their Jamaican cook who had often made me cinnamon toast when I was a child. I'd be propped up on a stool as she chattered about her life in Jamaica in her intriguing accent, a whole wall of refrigerators humming behind me.

Nearly every house on the street had a fleet of cars—an American car to drive downtown to work, a pair of foreign cars in the garage for the weekends, the wife's car, the nanny's car, the cook's car, and the gardener's truck. It seemed one simply could not have too many cars.

I looked across the golf course at the uneven line where the horizon met the sky. Objects were once again becoming themselves as the acid wore off, contained inside predictable borders, the colors of things once again muted. The leaves on the trees had stopped glistening like wet diamonds. Grass was grass; golf carts were golf carts.

I passed Henry Ford II's neo-Georgian estate where the Hugo Higbies had lived before him. Then I drove out of the gated entrance to our cul-de-sac, the armed guard waving me through from his mini brick fortress, and down one block to Lakeshore Drive.

I turned left, having no idea where I was headed. All I wanted was to drive along the expanse of the blue lake, listen to Fleetwood Mac, and smoke.

Peppering Lakeshore Drive—once the jewel of Grosse Pointe, with its sprawling lakefront estates built as summer houses in the first quarter of the century with automotive money—were scores of Mafia palaces. Most of the grand, old houses had been leveled, the properties subdivided. Few could afford the staff to run them. New houses had gone up on smaller lots. I studied their red-tile roofs, stucco walls, and flags of Italy flapping at full mast in the brisk wind off the lake. Mafia and automotive money now shared property lines, and everyone had bumper stickers pleading "Buy American."

The sun dipped low, setting the lake on fire, while above it a sky of blue glass began to crack with stars. I knew I should be home studying for my algebra test the next morning. But I'd had to get away from the sound of my father's voice.

I turned the car around by the yacht club and started heading back. The lights of Windsor were just coming on across the lake: Canada—our unlikely neighbor. My father had once taken me to Niagara Falls. We'd spent the afternoons in the wax museums looking at life-size wax replicas of famous actors, then on barstools in the saloons where I'd sipped Shirley Temples and feasted on salty peanuts while my father had cocktails.

Now my father's favorite bar was Gallagher's. Ever since Stroh's had expanded nationally by acquiring Schlitz and Schaefer, he'd begun to feel peripheral at work. I sometimes wondered if he'd been replaced without actually being fired. I knew he'd had no involvement in the new popular "Alex the dog" ad campaign, starring a golden retriever who fetched a can of Stroh's Beer from the fridge for his owner, then drank it himself.

Then, when my father's brother Gari was suddenly paralyzed by a riding accident, leaving him in a wheelchair with only slight movement of his neck, like Christopher Reeve, my father had declined even further. The two brothers had never seemed close, getting together only once or twice a year, but the accident had come as a huge blow to the entire family. Now I understood that my father felt things, deeper down, that he didn't let on about.

Each day at noon my father left his office to have lunch at Gallagher's and then camped out at the bar for the balance of the day. He came home in the evening as if he'd just left from work, wintergreen mints floating on his breath barely masking the scent of the booze.

Whitney and I usually stayed upstairs in our rooms during this witching hour and focused on our homework. Or at least pretended to. My father always entered through the side door, his footsteps falling heavily on the linoleum in the back hall, making my spine stiffen as I sat at my desk. Once he'd come into my room while I was writing a paper and had slapped me across the face, for no apparent reason. Later, he'd come back in, crying and apologizing. He was just drunk, he said.

The last of the evening light across the lake dissolved into black. There was nowhere else to go except home, though I wished I could just drive until I found the sun again, over the horizon beyond everything known or visible. I'd drive through wheat fields and cities and suburbs where other kids dreamed of leaving, and from there I'd keep going, never stopping until I came to that spot of sun I could sometimes see from this distance, or at least imagine was there.

. . . .

The dinner dishes had already been stacked in the sink when I came into the kitchen. My uneaten portion remained in a pot on the stove—Chef Boyardee canned ravioli. Jell-O brand chocolate pudding cooled in glass cups on the countertop. A salad of iceberg lettuce wilting in a wooden bowl. It was the kind of food we ate when my mother hadn't found time to grocery shop.

I took a fork from the drawer and speared the ravioli into my mouth straight from the pot. The pasta shell was still warm but the meat filling was stiff with cold. Sometimes my mother didn't heat things all the way through.

I heard the clatter of my mother's loafer heels on the dining room's wooden floor.

"Where have you been, Frances?" she said as she came into the kitchen.

"Studying with Andrea," I said automatically, still chewing over the stove. Andrea was my chemistry study partner, and I happened to be getting an A in chemistry.

"Andrea's a nice girl," my mother conceded. She went over to the sink and began rinsing the dishes, then loading them into the dishwasher. She wore a wool tweed skirt and a T-shirt with the word Bermuda across the front in pink script. The freckles on her calves danced chaotically as she sponged off the dishes. I watched the brownish spots with detached interest, hoping I could sleep in the aftermath of the acid. I planned to wake up at 4:30 a.m. to study for my math test.

I opened the refrigerator and stood staring at its contents.

A half-eaten honey-baked ham draped with foil, a macaroni-and-cheese casserole from the week before, three eggs.

My mother had either been at the real estate office (her new job), touring houses with clients, or playing backgammon. Both occupations had become her antidote to the bedlam at home while also helping to create it. And since she was the first woman in her family to work *and* to not have a cook, we were often left to forage.

"Is there anything else to eat?" I asked.

"I didn't have time to go to the store today," said my mother, switching on the disposal.

I was used to food being a last priority, but I was hungry. It had occurred to me more than once that if my father bought one fewer antique gun each year for his collection, we would surely be able to afford a cook, but the fact was, my parents didn't care much about good food. Perhaps in rebellion against their more formal upbringings, my father subsisted on Domino's pizza; my mother, on Campbell's soup and Saltines spread with Jif. My father hated sitting at the dining room table, and who could blame him? His mother had used the evening meal as an occasion to berate her husband and criticize her sons. As for my mother, she had been kept in the kitchen at mealtimes with the cook until she was nine years old, while the rest of the family was served meals in the dining room.

"Can I order a pizza?" I asked my mother. I was staring at the missing child's photo on the milk carton in the refrigerator. She was blond with home-cut bangs, maybe six years old. I wondered if they'd found her yet.

"I think Dad has some left over in the library," said my mother.

I was always hungry. By the time I was ten, I had learned how to make a good omelet, chocolate mousse, popovers, and pasta with bottled sauce. I missed the chickens Ollie would sometimes roast before she left to go home in the evening. But Ollie was gone now, back in Detroit and living on welfare, and with her had gone any sense of order.

"Ollie has to take care of her mother full-time now," my mother had told me.

I went into the library with a plate. My father sat in his leather chair, a remote in his hand, in his usual postwork outfit—khaki pants, dress shirt, and Topsiders with no socks. With his light-blue eyes and cleft chin, he looked like some famous actor whose name you couldn't quite remember. A Domino's box sat open on the floor at his feet with a half-moon of pepperoni pizza.

"Hi, Franny," he said with an absent smile. "Been studying?"

"Of course," I answered. My grades were good so my parents never hassled me about my whereabouts. "Test tomorrow."

I placed a slice on my plate. My father was watching *The Incredible Hulk*. Ever since the show had premiered, my father had co-opted Bill Bixby's famous line, "Don't make me angry. You wouldn't like me when I'm angry," offering up this threat whenever Whitney or I started to annoy him.

My father's photographs, painstakingly shot and printed, sat in piles throughout the room, still in their envelopes. Occasionally one landed in a frame, either at our house or someone else's, and the rest would eventually go into boxes, and the boxes into the attic. My father's eighty or so cameras cluttered the clothes and linen closets.

The tragedy of my father's life was contained in those dust-covered boxes. We hardly spoke about it, except when my mother would say, "Dad is so talented. He really should do more with his photography."

I'd grown up under the weight of all those unseen photographs and unused cameras, its burden so pervasive that even the air in our house seemed to have texture and mass. I observed my father on weekends in endless rounds of cleaning cameras and restacking print piles, or looking in vain for some shot of me on the terrace made the previous summer, to no avail. The mess just grew bigger, engulfing the pantry, then the kitchen, then the library, as my father was defeated by his own inability to edit, file, and catalogue.

Maybe the saddest story of my father's life was a missed meeting with his great idol, the renowned *Life* magazine photographer Alfred Eisenstaedt, who was scheduled to be on Martha's Vineyard in the summer of 1962, at the same time my parents planned to be there with Bobby and Charlie, just toddlers at the time. My father allegedly wrote Eisenstaedt a letter telling him he would like to meet. Taken with my father's enthusiasm for his work, Eisenstaedt responded that he'd meet my father at Gay Head beach; he named the date and time, but when the day came around, my father lost his nerve and stood his hero up.

The story was told and retold in our family, usually by my

mother, and soon took on the proportions of a mythical lost chance, coming to symbolize the tragedy of my father's unfulfilled potential. I always imagined Eisenstaedt standing alone on the beach at dawn, a tangle of Leica cameras around his neck, searching the coast for my father's lone figure in the mist.

Now my father was reviving his lost dreams through me. "You're a better photographer than I am," he liked to tell me. He loaded me up with equipment and encouraged me to photograph professionally, as he wished he had done. Since I was good at shooting people, I thought I might be a fashion or editorial photographer. I read *Rolling Stone* and *Interview, Vogue* and *Vanity Fair*. I studied the images, practiced the different styles. I was always experimenting. I felt as if I were tipping the scales of my father's lost chances by living the life he should have lived.

But my father's enthusiasm and support were something on which I could rely only when his mood was right. Other times, the house shook with his fury, and I turned to my camera, heading into Detroit or setting up shoots at friends' houses. The world through the lens was reduced to a manageable rectangle, and no chaos could penetrate the solace of the darkroom. Watching the clock while my prints developed and fixed lent predictability to my life, and I could cut a mat with the exactness of a surgeon. At the same time, my images were unpredictable, even mystifying; I never saw in my film what I thought I had through the viewfinder. I saw something realer, truer, as if separating a piece of the world from itself could somehow make the whole thing better.

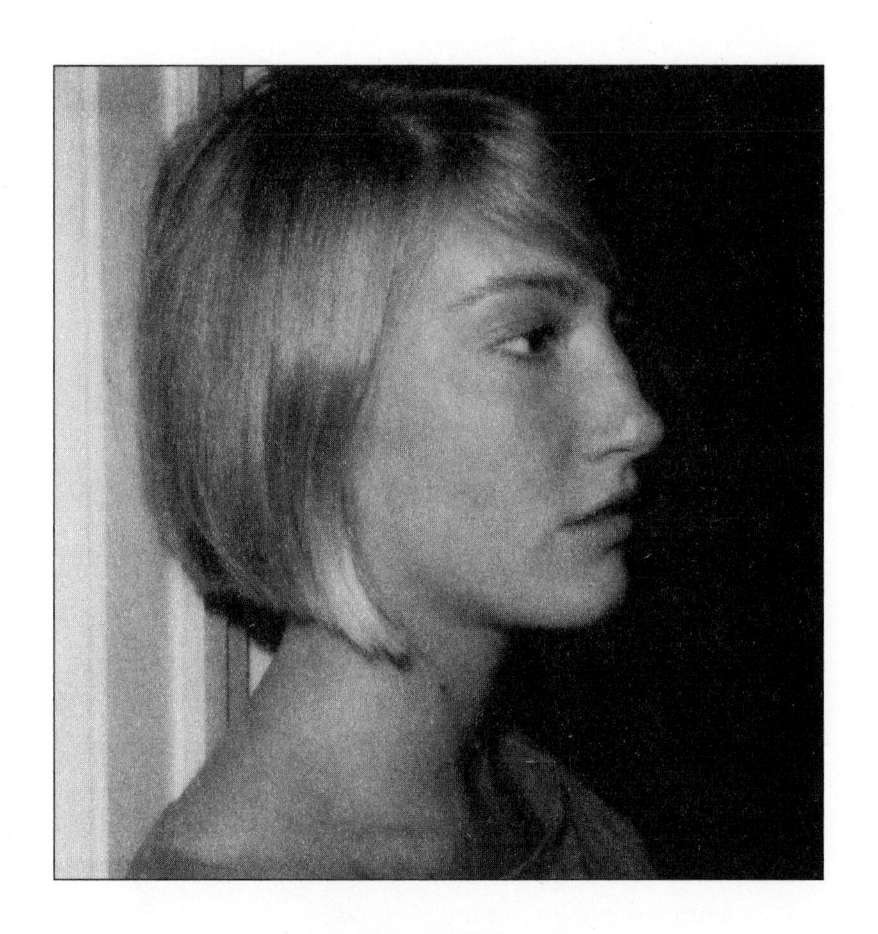

FRANCES STROH, 1984

(by Eric Stroh)

Whitney and I had been lying around the house all day, watching MTV and microwaving Stouffer's frozen French bread pizzas. The August air swelled with the metallic scent of an imminent thunderstorm. My father puttered around the house with his cameras, polishing lenses and blowing Dust-Off at me every time I walked past him in the kitchen.

He sat down and pulled a fish-eye lens out of its leather case. "Hi, Franny," he said warmly when I glanced over his shoulder. "Got another party tonight?"

It was the summer after my senior year, and I had been out every night. My father let me drive his Voyager minivan, even though I had totaled his Buick just a few months before, resulting in emergency surgery to eradicate the hematoma trapped inside the shattered cartilage of my nose. "It's Spence's birthday," I said.

Spence was my boyfriend that summer. The boyfriend who'd blown his entire summer-job salary to keep us both high on coke.

"Just don't wreck my car," my father said with a teasing smile. He reached out and tweaked my nose. He was in a good mood. "Make sure to avoid that fictitious dog, too, hunh?" He loved pointing out that he hadn't bought my alibi the previous winter that I'd swerved his silver Skylark sedan into a telephone pole—on the opposite side of the road—to avoid hitting a golden retriever.

"There *was* a dog," I said with a straight face.

"Right."

It was a dark morning in January that the wreck had happened. Partying our way through the inevitable depression of a Michigan winter, my friends and I had been out all night, then attended a sunrise meditation class at the Hare Krishna mansion in Detroit. The "Krishna center" was located on the sprawling estate of the Fisher Mansion, one of the old Detroit houses emblematic of the automotive industry's heyday, donated to the sect by Alfie Ford.

Meditation, music, drugs, and alcohol, they were all facets of the same mind-expanding trajectory—especially potent when combined. My friends and I had all read *On the Road* and *The Electric Kool-Aid Acid Test.* And with the help of state-of-the-art amphetamines and a healthy dose of cynicism we had taken the legacy of the fifties and the sixties to new heights—in the eighties.

The Corinthian pillars of the meditation hall were edged in gold. A robed, pot-bellied man with a Krishna ponytail sat lotus style facing the large group. We sat down in our stocking feet and tried to look spiritual. Sweet-smelling incense burned in all four corners of the vast room.

And now the chanting began. I glanced around at the other meditators as their voices rose. A beautiful woman with a shock of buzzed orange hair was sitting alongside us, cross-legged: *Annie Lennox*, her knee just inches from mine. Recognizing her immediately, we broke into ecstatic, wide-eyed smiles. We were living a cultural moment, absorbing her palpable aura of celebrity, metabolizing a cocktail of gorgeous chemicals, chanting "*Hare Rama, Hare Rama, Krishna, Krishna, Hare Hare . . .*" We had finally arrived. Annie was stunning, younger than we would have thought, and, amazingly, a real person. Her voice converged with ours like a train escalating to the heavens, echoing off the baroque, gold-leafed ceiling of the Fisher Mansion ballroom with the rapturous beat of life itself.

After the meditation class, my friends and I popped codeine tabs to soften the landing, then trudged through the snow back to my father's car.

The roads were thick with ice; the overcast January sky hung low like the dark concavity of an overturned bowl. I started the engine. Road conditions never worried me, even in blizzards. I could drive on ice blindfolded. I accelerated quickly, skidding against the curb.

"Whoa!" everyone shouted, laughing. They smoked and debated the age of Annie Lennox, seat belts still unbuckled.

The lawns were buried under filthy old snow. No other cars on the road. I accelerated again, feeling the pedal give obediently beneath the stiff leather sole of my right cowboy boot.

Brick houses whipped past us in blurs of reddish brown.

The car heater roared with cold air. I pulled a Marlboro from my pack. No one could find the lighter, so someone in the backseat just held out a lit cigarette. I turned around and leaned into the back, my left hand still on the wheel, my starved lungs drawing on that fragile point of light with mighty focus—the last burning ember within miles—but my cigarette didn't catch right away, and that's when it happened.

We slammed to a stop with a great exploding sound, our bodies thrown backward as if from an electric shock. Then everything stopped again.

A telephone pole, I saw, stood inches from my face, just beyond a windshield web of shattered glass. The front of the car was an accordion of crushed steel. I was still in the driver's seat. We were all still in our seats.

"SHIT," everyone said at once. We were alive, though.

My father put his lens back into the side pocket of his camera bag and buttoned it shut.

"What kind of dog did you say it was, again, Franny?" He asked me, still amused.

I picked up the Dust-Off and blew some air at the back of his head.

"Hey—Stop that!" he shouted good-humoredly. He glanced around. "Where the hell did your mother go, anyway?"

"No idea," I answered. My mother's mysterious absence was hardly unusual; she had been out even more since my father quit his marketing job at the brewery after a fight with my uncle Peter.

I poured a Coke over ice and went into the library, where

Whitney sat watching the music video of "Burning Down the House." David Byrne's absurdly blank expression bobbed around on the screen as flames consumed a suburban dwelling.

In the weeks that followed the car wreck, my parents hardly mentioned it. I had been expecting consequences, like being grounded from driving, but there weren't any. The huge splotches of blood they saw on my fisherman's sweater when they met me at the hospital—the steering wheel had saved my life, but done a number on me in the process—had subdued them. Thankfully, my friends had all walked away unharmed.

Then February brought the distraction of good news: I'd been accepted to Duke on early decision. My spotty career as a boarding-school-castoff-turned-public-school-lawn-urchin had finally ended; I'd made the grade. My parents were thrilled.

After that, I'd spent the last months of high school skipping classes more often than usual, drinking beer on the lawn of the War Memorial or getting stoned in the parking lot of Angel Park. I figured I'd earned it, and Mr. LeMieux, the assistant principal, still gave me a big hug at graduation when he handed me my diploma. South High was the first school where the administration and faculty had actually liked me. No matter how many of their rules I'd broken.

My mother's car pulled into the driveway just then and through the fishfly-covered screen door, we could hear her calling us outside. "Frances, Whitney, let's go for a drive!"

I turned off the TV, David Byrne's autistic face vanishing to some nihilism further yet.

We drove to the end of our cul-de-sac, and she parked next to the golf course. We never came down here; something was clearly wrong.

My mother didn't get out of the car. She wore a sleeveless blouse, Bermuda shorts, and her signature beat-up penny loafers. Her red hair had lost its luster, her once thickly lashed green eyes looked dull and lifeless, and her years of playing backgammon in the sun on the club's upper deck had left her badly freckled.

"Dad and I are getting divorced," she blurted out, still gripping the steering wheel.

I knew right away this must have been her idea. My father had been acting out for years—coming home drunk, yelling at all of us. In several fits of frustration I'd even told my mother to do this, divorce him; now that it was really happening, I felt a confusing mixture of shame and elation at the power I seemed to have over their lives.

I glanced down at her cottage-cheese thighs. She was fifty-one. How would she ever find someone else? I turned to look at Whitney in the backseat. His eyes were opened wide with shock as he scanned my face for a reaction. Neither of us spoke.

"But Dad doesn't know this yet," she continued. "My attorney is out of town for three weeks, and I can't tell Dad until he's back." She looked at us both sternly. "I just needed you to know so that you could adjust before school starts. So. Will you promise me, both of you, that you won't say anything to Dad?"

Whitney and I both looked down and muttered our agreement.

Satisfied, my mother turned the car around and took us back to the house, driving too quickly up the driveway, along the fringes of which my father had erected dozens of three-feet-tall metal reflector rods poking out of the grass. Ostensibly, the reflectors protected the lawn from tire marks. But this obstacle course only served to send my father into a rage each time someone knocked over a reflector on the trip up or down the driveway. Usually he would run out of the house to yell at the potential offender before it even happened.

The next three weeks I spent packing for college and observing my oblivious father with nagging waves of guilt and sadness. I knew he would be crushed when he finally got the news, but by then I'd be gone.

"Are you excited?" he asked me one day, standing in the doorway to my room as I sorted my cassettes. "Or are you going to miss us?"

I felt a sharp sting in my sinuses and suppressed the urge to cry. I slotted some U2 bootlegs into my cassette traveler. "Both, Dad," I said, my voice uneven, but we were talking about different things. The world he imagined I'd miss was already gone—he just didn't know it yet.

FENCE IN DETROIT, 1984

(by Frances Stroh)

When I arrived home for Christmas after my first semester at Duke, where I'd been learning to balance the enormous workload with nightly—and often all-night—parties, my father was living temporarily in a two-bedroom house on Mapleton Road, known as the "maids' road." My mother, who'd recently become a small-time Grosse Pointe real estate baron, had rented the house to him while she kept the big house on Provencal Road.

Christmas Day my father came over in a jacket and tie, as my mother's guest. Out of habit, he built the fire in the living room, welcomed the relatives at the door, took photographs of our group standing before the fireplace, the flames blazing behind us, but the tension of his dominance was gone, leaving in its wake an unsettling disorder.

By the fireplace, I watched Whitney set a glass down on an antique table without using a coaster. My father said nothing. Late the night before, after a party, I had come home long past curfew without any concern about my father's reaction. Bobby and Charlie talked loudly in the kitchen, drinking beer, after I'd gone to bed. And now our relatives' cars were parked haphazardly outside in the driveway—not in the neat lines my father had always enforced.

It was Christmas, my father's favorite day of the year, and everything was going to seed. He seemed distracted, smiling vaguely in the direction of laughter, changing lenses on his Leica more often than necessary. I wished I could comfort him. He appeared entirely unmoored.

"Let's get one more shot," he kept saying to the four of us. "Who knows when you'll all be home together again."

We lined up in front of the fireplace a third time. The heat burned into my back through my silk blouse. Charlie had recently been honorably discharged from the Marines and was sporting civilian clothes. Bobby and Whitney wore blue blazers and ties.

My father snapped the shutter. He lifted his cigarette from the ashtray and looked around the room for another background. The house felt empty without his guitars, his cameras, his mountains of pipes, books, and CDs.

The snow fell hard, thickly blanketing the roads within minutes of the snowplow's last round of exertions. People kept arriving at the house—Aunt Mard plunging her walker into the snow, Grandmother Susie in her mink coat and chinchilla hair—and dinner was postponed in favor of more cocktails. My father took coats and mixed drinks while my mother checked the turkey in the roaster. Bobby and Charlie appeared momentarily in the kitchen and beckoned me out and up the back stairs.

Bobby opened the door to his room. Three finely cut lines of cocaine sat in a row on the glass top of his desk.

"Here, Franny," Bobby said, handing me a rolled-up bill. "You go first."

A bird's wings fluttered inside my chest. My brothers had never offered me drugs. Bobby's mystifying room, where as a child I had spied on him kissing his girlfriend on the bed, where I had learned about sex from the porn magazines stored in the top drawer of his desk, had now become the place where the gap between me and my much older brothers—of age, gender, untold worlds of experience—would forever close.

Charlie sat on the bed smiling at me. His Marine crew cut hadn't yet grown out, underscoring the impossibly sharp angles of his face. He still lived in Southern California, where he was looking for a job. "Go ahead, Franny," he said encouragingly. "You'll like it. I promise."

Once upon a time Charlie's childhood room had been my refuge. "Sit down, Princess," he would say, patting the bed. And after we'd strung our bead necklaces, he'd tie mine behind my neck with kind, gentle fingers, as if not to break me.

I took the bill from Bobby and inhaled a line in one swift motion. "Wow," I said, feeling the searing heat spread through my sinuses. "That's intense."

"Don't worry," said Bobby. "That's the worst part. Just hold on."

I let them believe this was my first time.

Bobby did his line and Charlie followed. Bobby smeared his finger on the glass and rubbed the residue on his upper gums. He had a trim brown mustache and steely blue eyes that revealed no emotion whatsoever. We could hear our mother calling us from downstairs, but we lingered, pretending we hadn't heard her.

I sat down on the bed. The room was so cold I could see my breath.

Bobby leaned back in the swiveling desk chair. "Dad seems really out of it," he said. "You see how he'd light a cigarette, put it down in the ashtray, then light up another?"

"I saw him drink, like, four Cokes in an hour," said Charlie.

"Jesus Christ!" said Bobby. He licked his gums. "That man has a death wish." He unrolled the bill and slid it back into his wallet.

I felt a pang of protectiveness toward my father. All day he'd been going through the motions, trying to keep up appearances. "We can't imagine the hell he's been through in the last few months," I said. "Anyway, at least it wasn't four vodkas."

"He's just so screwed up," said Charlie, shaking his head, his face darkening with anger. "He pulled me aside earlier and told me the divorce was *my* fault. You believe it? Bunch of bullshit—everyone knows it was his drinking."

"Who the hell knows *what* it was . . ." Bobby trailed off, looking over at me.

I nodded silently. We all secretly believed that Charlie's coke bust had been the catalyst that led to my parents' marriage coming unraveled, if not exactly the cause. My parents had never known how to really talk to each other or comfort each other, and what little foundation they'd had simply crumbled apart. Scripted conversations weren't enough to sustain any real sense of connection. And so my father began drinking more, while my mother took refuge in real estate and backgammon tournaments, but the fact was they'd been headed their separate ways years before Charlie's bust ever slammed it all home.

That fall, after my mother filed for divorce, my father went into rehab, in hopes of getting my mother back, and a monumental weight had lifted from all of us when he became sober.

"Mom and Dad haven't gotten along in years," I added. "That's no one's fault."

Bobby looked over at Charlie. "Dad quitting his job didn't help."

"Whatever. He's the same old prick he's always been," said Charlie. "Mom did the right thing, getting out."

Everyone was gathered at the table when we came down, the coiffed, downy heads of great-aunts and grandmothers bobbing with the conversation. Plates had been piled with turkey, mashed potatoes, and French green beans.

My stomach turned at the mere sight of food. I sat down at my place, lifted my glass of wine, and took a generous sip. Everyone was talking at once, and I couldn't make out the words. The coke was speedy, the shaking of my wineglass just barely perceptible. I looked out the window and across the snow-covered lawn to the forest where I'd first smoked pot with some friends back in eighth grade. The blizzard had stopped, and the sun suddenly broke through the clouds. The surface of the new snow shone like splinters of shattered glass. I emptied my wineglass and poured myself another drink.

My uncles were droning on about our family's listing in the Forbes 400. I was aware of the fact, I guess, but I never really associated it with us. I'd bought a hundred-dollar bed for my college dorm room because the futon I wanted had been too expensive. *Forbes* must have done some interesting math if they thought we had that much money.

"The information in this issue is completely outdated," said Uncle Peter. "Fact is, the riverfront project's nearly *doubled* the business. Someone ought to write the editor a letter."

"Why don't *you* write the letter?" chimed in Aunt Nicole, her cropped blond hair tucked stylishly behind her ears, above a sleek black cashmere dress. "You *are* the CEO, aren't you? They need to get it right. It's ridiculous."

Uncle Peter took a bite of turkey with stuffing, chewed and swallowed, while the table waited for his response. "I'll write to them," he said finally. "Send them the plans for the Parke-Davis site."

It was a story I'd heard so many times. Uncle Peter and Coleman Young, Detroit's mayor, had shaken hands one windy day in a parking lot on the Detroit River next to the former Parke-Davis Pharmaceutical Company. The mayor promised Peter a stampede of tenants and greater visibility if the Strohs would develop the site, just upriver from where the Uniroyal Tire plant had been, and thereby improve Detroit's derelict riverfront. Coleman Young, who called himself the "MFIC"—Mother Fucker in Charge—certainly had the power to fulfill his promises, and Uncle Peter knew as much. Though personally I found it hard to believe anyone could ever revive Detroit. Seemed a little late to me.

Within a year of the famous handshake, the brewery had begun development on the site, launching a project that included a five-star hotel, a luxury three-hundred-unit apartment building, and a 1.8 million-square-foot office building. The development, renamed Stroh River Place when we'd bought the site, was the most ambitious development in Detroit since the seventies, when Henry Ford II built the Renaissance Center.

Soon afterward, Uncle Peter closed down the turn-of-the-century Detroit brewery and moved the corporate headquarters to Stroh River Place, saddening not a few Detroiters who took pride in Stroh's Beer being brewed in their hometown. The subsequent demolition of the historic building was even more demoralizing, but the family had already lost a hundred million dollars in production costs by keeping it open. Now the Stroh brand was produced in Allentown, Pennsylvania, in the newer Schaefer Brewing Company plant, where fire-brewing copper kettles had been installed, at enormous expense, to stay as true to our brand as possible.

"And send them that article from the *Detroit News*," suggested Nicole.

"Which article?" asked my father.

Nicole folded her arms across her chest. "Announcing that it's safe for whites to live in Detroit again. Because of the Strohs."

Uncle Peter chuckled. "The article didn't phrase it quite that way."

My father, who now got all his company news at gatherings like these, shifted restlessly in his chair. "Nonsense. No one gives a damn who lives in Detroit. No one's cared since the 1967 riots—'cept maybe the drug cartels."

Uncle Peter sipped his wine. "*We* do, Eric," he said.

Nicole turned to my father, superior. "We're in real estate now."

Everyone knew that Nicole basically ran the company behind the scenes, but no one was complaining; dividends had recently doubled.

"I know that, Nicole," said my father, controlling his temper. "I did work there until last summer." He made a point not to look at his brother.

My mother went into the kitchen to take the pumpkin pies out of the oven, glancing at Charlie's place on her way out of the room, monitoring his alcohol intake. Charlie ignored her. He reached over my grandmother's plate for the wine bottle.

My father sipped his glass of water and watched Charlie pour the wine. "Keep it up, Chas," he said bitterly. "You can handle it."

Seeing that my glass was empty, Charlie passed me the bottle across the table as soon as he'd filled his glass.

I poured gingerly, watching the sunlight pass through the carved crystal onto the white linen tablecloth to create floating fields of red. I wouldn't always be here, doing this; I wouldn't go down with the ship. For the moment, though, the wine warmed my insides and made the whole thing bearable. I picked up my fork and took my first bite of turkey, with no particular appetite for it, but it hardly mattered; the food had already grown cold.

Together

CHARLIE, WHITNEY, FRANCES, AND BOBBY—DALLAS, 1993

(by Cheryl Stroh)

Dallas, 1993

Charlie's condo complex sat just off the freeway, sandwiched between a strip mall and a sprawling warehouse advertising storage units for rent.

"Park over by the gate," Charlie said to Bobby, pointing to a chain-link fence with a pool on the other side. The condos were a forsaken collection of brown cardboard shoeboxes with mini balconies, each with its own laundry line.

"You live here?" asked Whitney from the backseat.

We all got out of Bobby's Volkswagen and followed Charlie through the gate and up a cracked concrete path.

It had been a little over a year since I'd seen Charlie, but he had the skin of an old hobo—mottled and bumpy, unshaved and scorched red, as if he'd been drinking moonshine under the blazing sun for the last forty years.

At thirty-two, he'd failed three rehab programs within five years. After a series of scenes at various family events, including

Bobby's wedding, my parents had banned him from coming home for holidays.

When I'd called about coming down to Dallas to film him, he was full of enthusiasm. "I love my family," he gushed. "You know? I really miss everyone."

My work had been selected as part of a group exhibition at San Francisco Camerawork Gallery entitled *The Family Seen*. Video screens of my family members talking would play in a darkened room simultaneously.

We'd all met in Dallas for a couple of days so I could shoot the interviews—the only time the four of us had ever met outside a family occasion. Whitney had come all the way from Missoula, where he was a senior at the University of Montana. I had come from San Francisco, where I lived and worked as an artist. Bobby and Charlie lived near each other in Dallas but hardly saw each other. I liked to think that art had brought us back together.

"Hold on a second," said Charlie. We followed him past a small play area with a warped plastic slide and a jungle gym. A group of people sat at the pool smoking, complaining in raspy voices about their "asshole bosses." Charlie leaned over the fence and greeted a shirtless man decorated with a collage of prison tattoos.

The man mumbled something to Charlie in a conspiratorial tone.

Charlie smiled. "Be over soon," he said.

Whitney gave me an anguished look and Bobby just rolled his eyes. Like my parents, Bobby had given up on Charlie long ago.

The freeway hummed behind us as we took the cement stairs to the second floor. Almost every door had a few pairs of well-worn flip-flops strewn outside. A bag of garbage sat leaking in the hallway. The earthy scent of pot smoke wafted out from someone's open door, and it occurred to me that this condo complex was likely the last stop, full of drifters like Charlie who'd lived in every condo complex in the Dallas area until this one, been kicked out each time for reckless behavior or failure to pay the rent.

Charlie opened the door. We stepped into a tiny living room that merged with a kitchenette. A beige sofa I recognized from the house on Grayton Road was the only furniture, other than a big Sony TV perched on some wine crates. The air smelled like sheets that hadn't been changed.

"Anyone want a beer?" asked Charlie, opening the fridge to a shelf full of Coors. I realized how proud he was to be hosting us and, though I never drank during the day, took a can.

"What's with the piss water?" asked Bobby.

"Got a case on special," said Charlie, the old yearning for approval audible in his voice; his big brother had finally come over to see him.

"No, thanks," Bobby said.

Bobby and Charlie's relationship had soured in recent years. When Charlie had a warrant out for his arrest in California for reckless driving three years before, Bobby had stepped in and gotten him a job at the Herman's Sporting Goods store he was managing then in Dallas. More than once, Charlie had come to work drunk, eventually losing his sales position, and their friendship never recovered. Since then Charlie had worked in a string of mini-markets shelving merchandise, and once as a gas station attendant.

Charlie took Bobby out onto the balcony while Whitney sat on the sofa and flipped through TV channels. Through the sliding glass door I could see my brothers talking the way they used to and wondered if, deep down, Bobby felt somehow responsible for Charlie. Had Bobby been the one to turn Charlie on to drugs in high school? I doubted it. All I knew was that the further Charlie fell, the more compelled Bobby seemed to feel to look away.

I went into the bedroom and attached the video camera to the tripod. I arranged a chair in a corner by the window. The bed was barely made, so I straightened the blanket to create a sense of order in the room.

Outside, the tattooed man held court at the pool. He smoked and talked and moved his arms wildly while everyone laughed, and I understood that all these people played a greater role in Charlie's life now than any of us did. All the missed holidays were adding up to Charlie's not really knowing us anymore. Our family was no safe harbor, but without us he'd been set, it seemed, irretrievably adrift.

"Is the camera on?" asked Charlie.

"Not yet," I said, as I adjusted the lens. Whenever he spoke, his left eye dipped into the viewfinder. "Try not to move your head." I hoped to keep his identity obscured by training the camera on the bottom half of his face. "All I want to see is your mouth."

I turned on the camera and went through my list of questions: *What was it like for you growing up in our house? How were you affected by Dad's drinking? Why do you think Mom and Dad got divorced?* and so on.

As he told his version of the family story, Charlie candidly discussed his coke bust in college, his years in the Marines, his downward spiral into drugs after his honorable discharge. "I made some bad mistakes," he admitted. "But that doesn't make me a bad person." His face began to sweat, probably from the stress of talking about the past, and I didn't want to go on for too long. I realized that while drugs had been a phase for me, they were a way of life for Charlie.

After I'd gone through my list, I paused. "What are you doing in Dallas?" I asked. It wasn't one of my scripted questions.

"I'm thirty-two and currently unemployed," he said to the camera, his lips spreading over yellowed teeth as he broke into a smile. "So let's hope the family business picks up and I can take early retirement along with everyone else." He nodded his head, still smiling.

I laughed. By "everyone else," I imagined he was referring to our father.

"So, that's your hope?" I asked.

"My number one hope and ambition is to come home for Christmas this year and see everybody," he said with a note of optimism. Then more defiantly, "*Mother*, I can come home and have just as much fun with my family not drinking as I can drinking with my friends."

He had rightly assumed that our mother would see the footage.

When we arrived at Ruby Tuesday, the bar buzzed with the local singles scene. A soccer game played silently on two enormous flat-screen TVs while Eddie Vedder's throaty, tortured voice bellowed over and over from the speakers, "Ohhh, ahhh, I'm still *alive*."

"Table's ready," Bobby said over the din.

The four of us walked into the restaurant, slid into a booth, and picked up the vinyl-coated menus.

With his Ralph Lauren–model looks, Whitney seemed utterly displaced in the Formica-trimmed booth. He glanced across the table at Bobby. "What do you recommend?"

"The potato skins," said Bobby. "With three-bean chili, highly recommend."

I studied the potato-skins offerings. "Do they pour the chili right on the potato skins?"

"Well, if you want it poured on you can get a twice-baked potato with chili."

"They also have great steaks," Charlie said. He'd had two beers at the bar during the twenty minutes we'd been waiting for the table. Bobby had told me privately he wished Charlie weren't coming to dinner. "He's guaranteed to get drunk and make a scene."

"He's coming," I'd insisted. "This is my one night in Dallas." I was leaving in the morning to film my parents back in Michigan.

We ordered the food and two pitchers of beer. The conversation floated from Bobby's vintage Volkswagen collection to his World War II uniform collection to the Stroh Brewery's poor sales record in Texas, where Bobby now worked as an area business manager for the family company, interfacing with wholesalers.

"It's pathetic," said Bobby. He took a long draw from his beer, leaving a mustache of foam on his mustache. "We have a brewery in Longview. We *make* beer in Texas, for God's sake, and we can't even sell our products here?"

"It's like the Busches not selling beer in St. Louis," said Whitney.

Bobby smeared his potato skins with sour cream. "Well, more like the Busches not selling beer in Colorado, but . . . I take your point."

"Maybe we shouldn't have gone national," I said. "You know, by buying Schlitz. I mean, don't you think the Detroit brewery and the Schaefer breweries in the East would have been enough?" I knew our troubles had begun when we got too big, when Great Uncle John could no longer walk the brewery floor, talking with his employees on a daily basis. There were just too many brewery floors now, and Uncle Peter was far less hands-on than Great Uncle John had been.

"No, Peter was right," said Bobby. "The industry was changing. Every other viable brewery had already made the move from regional to national. It's just, you know, that we grew too *fast*. We were underprepared. There was no strategy other than just to grow for the sake of it. And we borrowed too much to finance the Schlitz deal, of course. That's what's killing us."

"Someone could have bought us out," I said. "We must have been worth . . . *something*."

"Not enough," said Bobby. "Lot of mouths to feed in this family."

"How many brands do we make, anyway?" asked Charlie. "I can't keep up."

Bobby dipped his potato skin into his chili. "About thirty?"

Whitney cut into his steak. "And are *any* of 'em doing well?"

"Not in the U.S.," I said, through a mouth full of French fries. "Far as I can tell . . ."

"They have Stroh's on tap at the bar here," said Charlie. "So . . . sales can't be *that* bad."

I slid past a row of four knees en route to my window seat. I took a Russian novel out of my bag and began the five-hour dissociative state that the flight to Detroit always called for.

I put the book down and gazed out the window. They were loading the luggage onto the conveyor belt. My camera equipment sat snugly in the overhead compartment, the tapes stored in the foil-lined bags my father had given me at Christmastime. "Never forget to put your film in here," he'd said, handing me the bags unwrapped. "X-ray'll destroy everything."

He still thought of me as a photographer. "Why don't you keep shooting pictures?" he'd asked when my work had taken a new direction just after college. "You're a damn good photographer."

But I'd begun to feel limited working in two dimensions and had a feeling that becoming a master printer wasn't in the cards. I didn't have the patience. The truth was, I suspected I wasn't very good.

I took classes at the San Francisco Art Institute after Duke, mostly in their New Genres Department, and there I entered the world of video, installation, and performance. I began to think in terms of narrative and space, context and concept, signs and signifiers. Later I joined the MFA program at Art Center in Pasadena but found that the making of art there was essentially peripheral to the reading and discussion of French critical theory.

As for my loft in downtown Los Angeles, with packs of wild dogs in the streets and Interstate 10 practically grazing my windows, it was a lot like a scene out of J. G. Ballard's *Crash*. When the L.A. riots broke out after the Rodney King verdict, I could see seven different buildings burning within a mile, not to mention machine-gun armed militia roaming the streets of my neighborhood. L.A. felt like the end of the earth; and I missed San Francisco.

I dropped out after my first year at Art Center. The program felt soulless and the art derivative—my own in particular. As a farewell, I did a site-specific piece in the school's Bauhaus-style building by silk-screening over the words *fire extinguisher* with *soul extinguisher* on the rectangular black extinguisher boxes lining the hallways. My boyfriend Marko—also an artist—pulled the squeegee while I'd held the screen.

Now I was back in San Francisco and single, with a well-paying job at an interior design firm and several shows of my installations scheduled over the next year, the family piece being the first. So far, my family members were cooperating. In fact, the combination of lights, a camera, and a list of open-ended questions seemed to open a Pandora's box of responses that I never could have anticipated; everyone seemed to have a pent-up need to talk about the family.

As practice, I'd been experimenting with friends in San Francisco, filming their answers to questions, then editing together only the responses, one cut after another. The stream-of-consciousness, solipsistic effect was powerful, rather like the way it felt to be privy to interior thoughts of one of Tolstoy's characters.

"How's Charlie doing?" my father asked. Seated in his leather armchair watching a Western, he pointed a Colt 45 revolver at the TV screen every time John Wayne pulled his six-shooter from his holster. My mother had once told me of a similar scene when Charlie's baby nurse came downstairs to the library. "Mr. Stroh," the poor woman announced, standing at the entrance. "Charles has taken his bottle." Sitting there in full Western regalia, replete with a cowboy hat, chaps, boots with spurs, leather holsters, and a revolver in each hand, my father kept his eyes on the TV. "Thank you, Ivy," he said.

I attached my video camera to the tripod, aiming the lens

at my father. "Charlie wants to come home for Christmas," I told him.

My father frowned. "He drinking during the day, or only at night?"

"Both," I said. I moved the tripod and adjusted the height so that the viewfinder framed the bottom half of his face. "I'm worried about him."

"Is he working?"

"Not right now."

My father switched off the movie with a remote and set the gun down on a table. "You'll have to talk with your mother about Christmas," he said. "That's her department."

I scanned my father's library for additional lights with which to better illuminate his face, but none were quite the right height. I walked through his entryway and down two steps into the sunken living room. The lushness of the garden flooded in through the picture windows. Out on the terrace, large ceramic ashtrays shaped like fish had been placed on glass-surfaced tables. Remembering these brightly painted ashtrays from the garden parties where, as a child, I often finished the cocktails left behind, I marveled that they had lasted all these years, through so many parties, so many moves.

My father had settled into this six-bedroom house five years before this. Looking around at his immaculately decorated rooms, a visitor could have been forgiven for assuming that our beer brands were thriving. But Uncle Peter's late entry into the exploding light-beer market in the eighties had kept us from competing effectively. In 1989, after a deal to sell our business to Coors had fallen through, dividends had been withheld, and the family had started attending annual shareholder meetings—something we'd never done—to be told how poorly the company was performing. Angered by our advertising cuts, wholesalers gave up on our brands, switching to rival labels, causing sales to drop precipitously. A sudden, panicked repackaging of the iconic Stroh's brand—to block letter blue, contributing to a 40 percent sales decline within a single year—was the last nail in that brand's coffin. The new packaging might have worked, had we not tried to market the very same beer formula as a higher-priced premium beer. As it turned out, my father hadn't saved a thing, and my mother had to lend him his mortgage money until dividends were resumed a year later.

In 1990, in the aftermath of the failed Coors deal, management rallied to revive Old Milwaukee, our strongest brand, by committing an unprecedented $9 million to a new marketing blitz: the Swedish Bikini Team ad campaign. The celebrated Hal Riney Agency in San Francisco came up with the concept of svelte blond-wigged babes with cases of beer dropping out of the sky in parachutes, or coming downriver in a boat, to update the already popular "It Doesn't Get Any Better Than This" Old Milwaukee ads. For the sake of variety, viewers saw a different Swedish Bikini Team commercial on TV every night. Many considered it the best beer campaign ever made. Ratings were off the charts.

Then the bomb dropped. A group of female workers in the Stroh's plant in St. Paul, Minnesota, hired a powerful feminist lawyer, claiming sexual harassment and discrimination in the workplace and linking it to the ad campaign. The controversy sparked a national press debate over sexist themes in advertising. The fact that the Swedish Bikini Team had also just posed for the cover of *Playboy* magazine didn't help. Feminist groups everywhere used the campaign as a scapegoat. Soon the media controversy landed on the cover of the *Wall Street Journal*, and the Minnesota lawsuit was headed for the Supreme Court.

Stroh's finally pulled the campaign in its third month, spending minimally on advertising from that point on. It was impossible to compete with Anheuser-Busch's deep advertising pockets. And so management switched to a strategy of managing our U.S. business down while making a big overseas push, particularly in India and Russia, where Stroh's Beer was fast becoming an icon of the American way of life.

Out on the terrace a light rain began to fall, pelting the ceramic fish ashtrays. I pulled a plug out from the wall and carried a lamp from the living room into the library.

"I'm thinking of selling this house," said my father.

I put the lamp down. "Already?"

"Business isn't getting any better." He helped me plug in the lamp under a table stacked with photography magazines. "And our real estate's in the damn toilet."

The riverfront development, he meant. Businesses were leaving the city in droves. We'd already lost the apartment building and the hotel to foreclosures. "It's amazing," I said. "General *Motors* is making cars in Mexico, and somehow the Strohs are still stuck in Detroit."

"Damn right," said my father. "And instead of advertising our core brands, your uncle Peter throws tens of millions into biotechnology. Nice move, hunh?"

On a recent family business weekend in Durham, North Carolina, I'd toured the impressive research center we had built in the Research Triangle Park. We were developing a drug, apparently, to treat septic shock, but I knew the likely success rate of such a venture was close to zero. "More money down the drain," I agreed.

"We've been de-listed at *Forbes*," added my father, his voice taut. "We'll be broke before we know it."

His stress was contagious—I could feel my own heart rate picking up. I tried to calm my breathing as I adjusted the light. My mother often worried about the fact that I was an artist; she still held out hope I would go get an MBA and get serious about my life, make some money, and drop all this art silliness. The chronically bad company news made me wonder suddenly if she was right.

"We're headed for a big fall," my father said with an air of finality. He sat back down in his chair and lit his pipe, drawing on the smoke with rhythmic breaths to get the embers going.

He'd been sober eight years now and seemed settled in his life, often seeing my mother for dinner and even traveling with her on occasion. In many ways, they still operated as a couple, my father maintaining my mother's car, my mother doing his laundry. I found their continuing friendship as reassuring as the bad company news was persistently alarming, and I know my father did as well. After their divorce, my parents had managed to brick together another foundation that seemed to bolster all of us—or at least Whitney, Bobby, and me. As long as those bricks held firm, I imagined we could live our lives productively and with resilience, even if the company itself fell apart.

I looked around at the room. Leather furniture with nail-head trim, antique chests of drawers, valuable volumes on the bookshelves, blackamoor figures flanking the fireplace mantel. How could we have been so clueless? Sexism and racism seemed to permeate everything, especially our beer ads. I'd watched our Schlitz Malt Liquor and St. Ides commercials targeted at inner-city African Americans and felt embarrassed about the stereotyping. These ads weren't helping sales either.

No question, we were headed for a big fall. I imagined the repo man loading everything in my father's library into a truck and driving away. The money had never been mine and probably never would be, yet the fear was as familiar as air; we'd always been on the precipice.

"Oh, well, enjoy it while it lasts, Dad," I said, turning on the camera. "'It doesn't get any better than this, right?'"

My father smiled and exhaled a cloud of smoke.

SOUL EXTINGUISHER, 1992

(by Frances Stroh)

On a hot, windless day that everyone called "earthquake weather," I sat in an editing room at the San Francisco Art Institute and pieced together my family's answers to my interview questions, trying to imagine the psychological effect on the viewer when all six narratives overlapped. I knew the visual effect of six talking mouths in one room would be captivating. In the tapes, Whitney's and Charlie's veneers cracked quickly, their upper lips beading with sweat, voices leaking bitterness. Bobby and I were subdued, philosophical, bent on humor. My parents were distant and instructive, looking down on their children from lofty heights, particularly when talking about Charlie.

"The problem with Charlie is he never grew up." My father paused and puffed on his pipe. "We wish it were different, but it isn't. If you asked him today, he'd deny that he has any problems with drinking and drugs. He'd say, 'Oh *no, I* don't have a problem.' "

My father went on to share an exercise he himself had learned in rehab. "You make a list of things you like about yourself in one column, and in another column you list the things you don't like. Then you study the list and figure out what you can change about the things you *don't* like. Some things you can't change, but most things you can." He drew on his pipe with a certain satisfaction, talked about how he "took the cure," then went on to admit that he still drank while on planes.

"Dad's drinking was the catalyst that led to the divorce," said Bobby. "And Charlie's problems may have surfaced during a fragile time in their marriage, you know?—which only caused Dad to drink more."

"There's a myth in my family that Charlie's alcoholism and drug addiction caused the breakup of my parents' marriage," I said.

"Charlie is a junkie, an alcoholic," said Whitney. "He's stolen from his family, he's lied, he's cheated, he's shat all over them. It's broken their hearts."

"Dad blames me for the divorce," said Charlie. "But that's a bunch of bullshit. It was his drinking."

"Each of my children has been deeply affected by Charlie's problems," said my mother, "but in different ways. My ex-husband's been the most affected, to the point where—to the point where he can hardly talk about it." She went on to say that my father had been ineffective at his job at the brewery, and that her sons were "procrastinators."

"But I see myself in my daughter. She is striving to achieve, as I am in my own life. I think if I had those years of mothering to do over again it would turn out just the same. I'd like to have control over my children's poor decisions, but I find I don't."

As I sat in the darkened editing room, something unexpected happened, something for which I was not prepared: I had my first conscious glimpse of just how profoundly screwed up we were, how detached, inhuman, even. Each of us discussed Charlie as if he were nothing more than a character in a novel we'd recently read. We had learned well the art of detachment, my siblings and I, from our parents and perhaps theirs before them. I had been skating along the surface of the pain for years, pretending to look deeper, outwardly dismissing my parents' version of reality, even rebelling against it, all the while inwardly accepting it as fact.

Our family was like one of those hand-painted road signs that point in a multitude of directions at once: laziness and bad genes were the problem, according to my mother; according to my father and Whitney, Charlie himself was the problem; Charlie would have it that our father alone was the problem; while, according to Bobby and me, an unfortunate alchemy of both Charlie's and our father's problems was to blame.

The cognitive dissonance between my parents' versions of the story and ours simply could not be reconciled. I had written a paper to be presented on a panel at the gallery discussing my piece in purely conceptual terms, yet now I was unearthing a truth that could not be bound by any intellectual discussion. Looking at the piece as an outsider, I liked the tension of the raw emotional material pressing up against the cool, minimalist look I'd chosen—those six rectangular screens displaying enormous talking mouths—but these had nothing to do with me, with what went on inside of me when I myself watched the tapes: the horror, the shock of recognition. The emotional foundation I'd imagined was there, I realized, had been cobbled together out of repression and denial. Time was passing, things were falling apart, and we seemed unable to catch ourselves.

Switching off the light in the editing suite, I pulled the door shut. When I came outside into the warm dusk, the sculpture students were piling debris into the parking lot Dumpsters. Clouds of white dust rose up into the air every time they hurled something in, and I had the sudden impression that art was nothing more than a gratuitous accumulation of clutter that would someday have to be thrown away. The sight of discarded ceramic heads, welded metal shapes, and cracked wooden pedestals heaped into the trash was demoralizing, and I briefly wondered if my work, too, would amount to this sort of "nothing," the image of my father's stacks of photographs flickering through my mind.

I walked down the hill on Chestnut to Columbus and up toward City Lights Bookstore. The street hummed with life, people sitting at outdoor tables, crowded Italian cafés. I crossed the street to Caffe Roma, taking the last table outside.

I gazed across the street at the Art Nouveau stained glass sign of Vesuvio Café, where the Beats had congregated back in the fifties—Allen Ginsberg, Jack Kerouac, Lawrence Ferlinghetti, and Gary Snyder. Suddenly I found myself recalling an evening I'd spent with Allen Ginsberg and William Burroughs during my senior year at Duke. It was one of the most memorable of my life, that evening, though I rarely told anyone about it, lest I be expected to share some revelation of cosmic proportions, when in truth the conversation, and the poetry reading that followed, had been disappointing, certainly, as compared with the magnitude of their celebrity.

Like so many college students in the eighties, I'd been fixated on the counterculture of the fifties and sixties—from Andy Warhol and the Velvet Underground to Bob Dylan and the Beats. A religion major with a focus on Buddhism, I was headed to San Francisco as soon as I graduated, with high hopes of establishing myself as an artist there. I'd read all the Beat writers and identified with their antiestablishment ethos; as a psychology minor, I'd been influenced by the writings of Aldous Huxley, Timothy Leary, and Alan Watts, which had led to my research paper on the clinical use of LSD in the treatment of alcoholism. The local Hare Krishna chapter had been the subject of my photo essay for a documentary photography course; I'd even considered having the Hare Krishnas cook one of their famous vegetarian dinners to honor Ginsberg and Burroughs.

Instead, I organized the dinner at a local vegetarian restaurant before Ginsberg's reading. At a table of twenty people, mostly Ginsberg's posse, along with a few of my friends, I was the only woman. But in spite of the Beats' notorious indifference toward my gender, I'd managed to seat myself next to Ginsberg and directly across from Burroughs. Tommy, my four-year college boyfriend, with whom I'd finally disowned my virginity freshman year, was also seated next to me. A clean-cut prep school kid when we'd first met, he now sported shoulder-length hair and a beard, while my own hair hung down to my elbows, crowned by my signature Greek fisherman's cap. We sat smoking Lucky Strikes, too shy to speak, completely awed by our dinner guests.

Ginsberg, with his wandering eye, was jovial and troll-like as he held court. Burroughs sat regally and silently in his dinner coat, lighting one cigarette after another with a dramatic sweep of his arm each time he struck the match. I'd half expected to meet the two youthful men I'd seen posed with Paul Bowles in a photograph, *Tangier, 1961.* But these men were old, Burroughs in particular.

Ginsberg showered Burroughs with doting attention and private jokes that made the rest of us feel like onlookers. They discussed the Lemur Center at Duke—one of the reasons Burroughs had desired to make the trip to Durham. He was a big fan, apparently, of lemurs.

When the salads arrived, Ginsberg praised the restaurant's tofu-tahini dressing.

I seized my chance to join the conversation. "I know. I'll miss this dressing when I'm in San Francisco," I told him.

"Ah, San Francisco," said Ginsberg with an ironic little twirl of his salad fork. "And what 'ill you be doing there?"

"I'm going to be an artist."

Nodding vaguely, he encouraged me to spend time at the San Francisco Art Institute. "It's a wonderful place," he said. "I taught a writing workshop there once."

After dinner we all drove to an auditorium on campus where Ginsberg would read. My friends and I sat on the edge of the stage, drinking red wine from the bottle, as the beatniks had once done at Ginsberg's quixotic first reading of "Howl" in North Beach. But the mood wasn't there, the sterile auditorium half empty. Ginsberg read a series of sexually explicit homoerotic poems, and people started filing out. His provocation of the mostly conservative-looking crowd appeared intentional, and perfectly in character; he read one lewd poem after another, a determined smile on his face, clearly getting a kick out of the audience's reaction. Toward the end, he read "Sunflower Sutra" and a string of older work, rewarding those of us who'd stayed.

Afterward, we went into a brightly lit reception area with white tablecloths, wine, and platters of cheese. Ginsberg came up to me with his big grin. "How did you like the sphincter poems, Frances?"

I sensed he was making fun of me because I'd been the only woman at dinner, surrounded by male sphincters. "I thought they were great, Allen," I replied witlessly. That was the power of the icon that was Allen Ginsberg: he'd lost the majority of his audience, and yet I would be the one to obsess over my unclever reply to him for years to come.

It didn't matter that most everyone had walked out of the reading, or that I hadn't known how to reply to Allen Ginsberg, or that William Burroughs couldn't have been bothered to say a word to me or anyone else at dinner. They were two of the most legendary figures of the twentieth century, and I had spent an evening with them. I had made an important discovery, too: they were only people. We all were. And some of us were also artists, or trying to be artists. Everything was happening on a continuum, I saw, and I allowed myself to believe that night that maybe I would be next, that perhaps a bit of magic dust had floated my way, off the stiff shoulders of these two old men.

I still had the signed copy of Ginsberg's *Collected Poems* on my bookshelf in the Haight. It had survived the '89 quake, when my bookshelf toppled over, crashing into my dresser, sending my books in every direction. Ginsberg had signed the book, his childish inscription the only evidence that still remained of my evening with the Beats.

San Francisco Camerawork Gallery was located just south of Market Street in a former warehouse boasting floor-to-ceiling windows, creaky hardwood floors, and a freight elevator that could carry a crowd, twenty at a time, two floors up to the opening reception.

I'd spent nearly twenty-four hours installing my piece, breaking for a short nap on the gallery couch sometime before dawn, and afterward heading to the airport to retrieve my parents. Now my mother stood paralyzed in my installation room while my father peeked in from the door, as if afraid to enter.

I picked up a plastic cup of wine and took in the rest of the exhibition, mostly innovative photographs of the artists' interpretations of the show's theme, which was family. On one wall, enormous color murals of family beach scenes offset a black-and-white triptych of an African American family posed in front of their church. My piece was the only installation work.

"Good work, Franny," said Anthony Aziz, the gallery board member who had proposed my piece for the show. "Powerful stuff. I spent thirty minutes with your piece—longest I've ever *spent* with a video installation."

I'd met Anthony when he was a graduate student in photography at the San Francisco Art Institute. Later, he'd teamed up with Sammy Cucher, a friend of mine from the New Genres Department, to form the collaborative artist team Aziz+Cucher.

Glancing across the crowded gallery, I thanked Anthony for including me in the show. I noticed a line forming at the entrance to my room. "I think my parents are hogging all the space," I said. "God, maybe I should take them out for dinner."

Just then, Anthony, always so well groomed and composed, shot an alarmed look in the direction of my room. "You're kidding, right?" He stared at me. "Your *parents* are here? How are they . . . going to, you know, *react*?"

I had the sudden sensation of waking up out of a dream. After so much insight in the editing suite, I'd somehow gone completely numb again in those remaining weeks before the exhibition, focused as I was on the formal and technical aspects of the piece. I'd done an even cleaner edit and hired a technical assistant to help me with the wiring of the piece in the gallery. Alone in the editing room, I'd been overcome with horror at my family's collective dissociation from its own demise, only then to end up feeling nothing at all. I'd forgotten that stark epiphany, bottled it up as I'd always done, creating an emotional debt to be paid later.

"It's nothing they don't already know," I told Anthony, more evenly than I felt now. "They lived it, right?"

He gave me a concerned look. "It's just . . . your brothers' monologues are so honest, you know? Everyone makes statements you'd imagine might be . . . just a little too painful for the others to hear."

"I guess that's my role in the family: can opener for the can of worms!" But I wondered if I'd opened something I shouldn't have.

Anthony laughed and put his hand on my shoulder. "Anyway, good luck," he said. "And keep going, you hear me?"

I listened to every tape until it repeated," said my mother. "I stood right next to the TV screens so I could hear everything."

"Not *me*," said my father gruffly. "Stick with photography, Frances. That's *my* advice."

My parents were staying at the Stanyan Park Hotel in the Haight, just across from Golden Gate Park. It was one of the many ways in which they still operated as a couple, traveling together, staying at the same hotel.

I took Fell Street from Zuni, where we'd had dinner, up to the Haight. A crowd had gathered in the Panhandle, encircling a string of bongo players as they danced wildly to the discordant rhythms. With all the whirling tie-dye and dreadlocks, the scene had the feel of a Grateful Dead concert. I turned to my mother in the front seat. "Did you hear Charlie say he wants to come home for Christmas this year?"

"I heard that," she said tentatively. "He looks like the devil, though, doesn't he? He's not taking good care of himself."

"He's a goddamn drunk!" trumpeted my father from the backseat.

I turned left onto Stanyan Street. The cherry-blossom trees in the park were flowering, the grass underneath them covered in pale-pink petals. The drifters and street people had gathered into clusters, some wrapped in old blankets, talking amiably in the last of the April light. The rough, red skin of these homeless people reminded me, of course, of Charlie.

"It's his last stop," I said.

"Whose?" asked my mother.

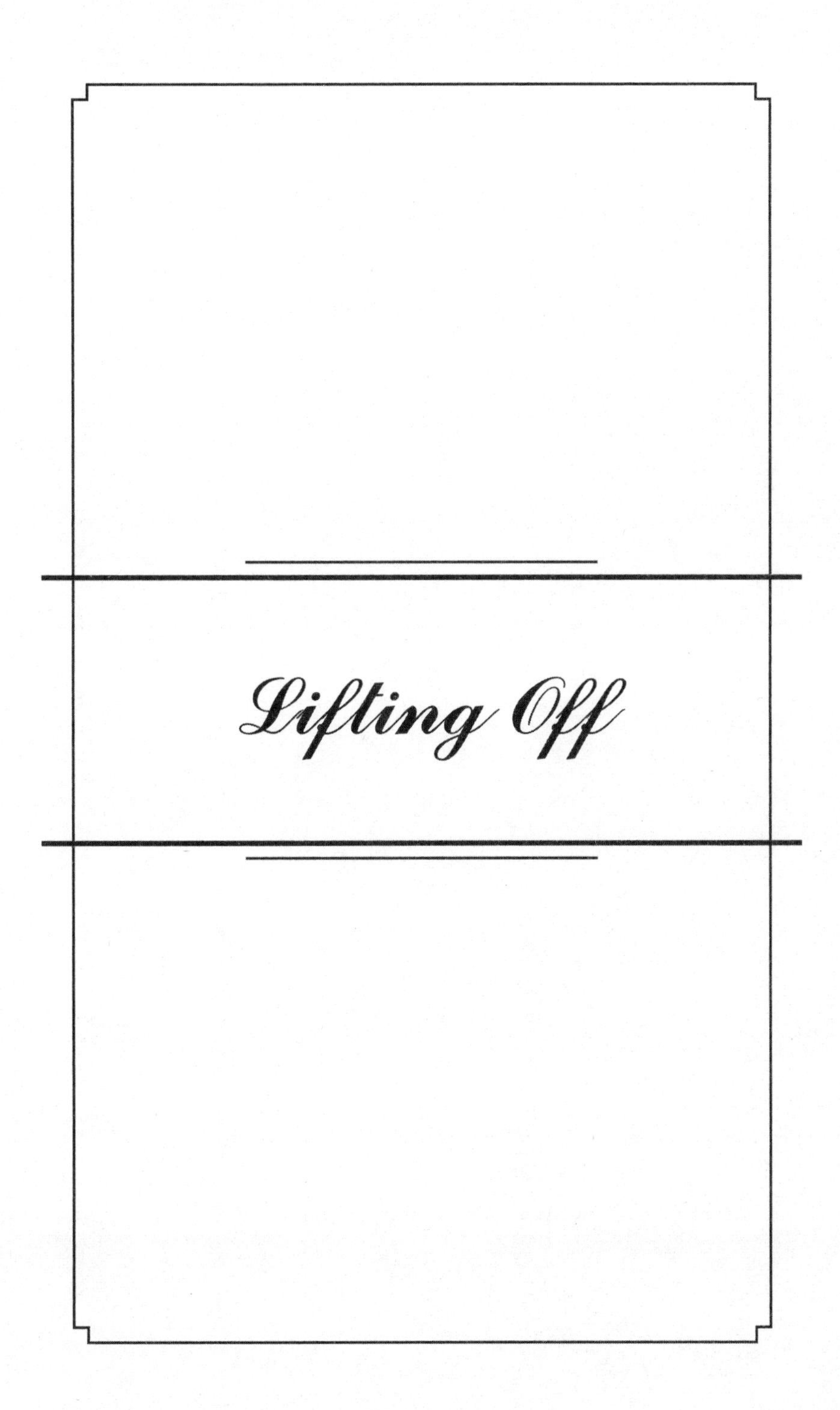

Lifting Off

***FRISK BOTTLE*, 1995**

(by Frances Stroh)

London, 1995

My friends and I walked up the King's Road in the light September rain, headed for the 606 Club. We'd been drinking martinis at my new flat as a housewarming, the gin still dry on my tongue. I savored the rain pelting us, pelting the pavement.

"It's down this way," I said, turning onto Lots Road. Smoke-stained brick warehouses ran all the way down to the Thames. A crowd had gathered in front of an unmarked building. Supper club jazz floated up the stairs from the basement. We got in line.

"I never knew about this place," said Hari. Charismatic, with movie-star looks, Hari was studying acting in London for a couple of years. We'd been friends back in San Francisco.

"It's a hidden gem," said my new friend Nino, paying for the tickets. He wore slicked back hair and a boxy suit from the forties. "Members only." We'd met at Camden Market, where Nino sold me a vintage dining table and chairs for my flat.

Camilla, Hari's tipsy red-haired girlfriend, slid her arm through his and smiled. "It's *lovely*," she said.

I'd been out every night for a solid month since arriving in August, raging on adrenaline ever since I'd been awarded a Fulbright for a year of study in London to complete my Master of Fine Arts. With the ascension of the Young British Artists—stars like Damien Hirst and Tracey Emin—London was the epicenter of the art world. Some days I had Imposter Syndrome, unable to believe my good fortune. Then I reminded myself I was here to begin my life anew, put the past behind, and launch my career as an international artist. This was where I would make my name. Who knew? I might never return to the States.

Inside the smoke haze of the 606 Club, we were given a table near the stage. The keyboardist jammed while the female vocalist sang with a voice rather like Billie Holiday's. The guitarist hummed the chords as he played. Nino ordered a round of drinks.

"Watch the guitarist closely," he said seductively into my ear. "It's like he's making love to those strings, right?"

Nino and I only talked about sex. We'd been to swing dance clubs and martini bars all over London, but we were just friends. After a series of failed relationships in the States with men who'd been remarkably like my father—addicted, adoring, creative, yet essentially self-absorbed—I'd promised myself to stay single and focused on work for a year. Relationships only diffused my focus; months would pass, my attention fixed on the inevitable ebb and flow of closeness, my ideas rattling around in my head rather than taking concrete shape, my creativity sapped. I was twenty-eight, and it all seemed such a chaotic time, those years I'd spent in San Francisco dating and having relationships, fitting in exhibitions here and there while trying to figure out how to make a living and still be an artist. Now, in London, looking ahead into the glaring light of my bright future, I was finally on the right path, and I wasn't going to let anything get in my way.

Hari and Camilla sipped their drinks. They'd met in acting school. Back at my flat on Wandsworth Bridge Road, we'd sat in my conservatory, with the rain hitting the glass roof, pouring one drink after another. In the States I could hardly drink without feeling ill, but here it seemed I could drink as much as I wanted. It felt as if I'd somehow outrun myself.

Hari smiled over at me and tipped his drink back. I felt lucky to have such a good friend in London. We shared an extended network of friends back in San Francisco, the closest thing to family some of us had ever known. Everywhere I'd ever lived, my friends had been my surrogate family, but nowhere more than in San Francisco. Our crowd threw Mexican "family dinners" every Sunday night, with platters of catfish and black beans and tortillas—a tradition that Hari and I had decided to continue with our London friends.

"Tomorrow," said Hari, "we're going to get you a bike."

A bike would help; I didn't own a car. Only I was nervous about navigating London's fitful traffic, riding on the opposite side of the road. "I won't have to wear one of those masks for the fumes, will I?"

Camilla laughed. "With that amazing hair?" she said. "A crime!"

In celebration of my new expatriate identity, I'd cut off my long, blond hair in favor of a layered sixties bob. Together with the thick black eyeliner, fake lashes, and the high, lace-up boots I'd adopted, the effect was a throwback to Andy Warhol's silver foil-lined Factory. Astride my new Raleigh three-speed, I'd be a cross between Edie Sedgwick and some rave-babe bike messenger.

Nino took my hand and led me onto the crowded dance floor, pulling me in close, his warm hands on my waist. I caught my reflection in the tilted mirror behind the stage—watching that London hipster with the shock of platinum hair groove—and thought for a moment I'd spotted someone else entirely.

It felt so liberating, leaving behind my family and the failing business, as if I'd shed a too-tight suit and could at last move freely. Walking down a London street, I sometimes imagined myself in the final jailbreak scene of *Midnight Express*, feeling as if at any moment I might bolt into a run, my sense of buoyancy too much to contain.

As if to test my resolve, my father had begun appearing in public with a young woman named Elisa Keys. He called me in London to tell me that his new girlfriend and I had attended Grosse Pointe South High School together. He seemed proud of this.

Not recognizing her name, I couldn't recall her face.

I responded as I did with all bad family news. First I felt an intense wave of panic, then shut it out. It was a distant storm. My family's ship might be sinking; my own, though, was just setting sail.

One day my cousin John, who would soon be the new CEO of the family company, called me, fraught. "Franny, you have to *do* something about your father and that woman."

Before my father married her, I knew he meant.

"John, what *can* I do?"

While John worried about our family's public image, I ruminated about my father's health. I heard my father was drinking again, after eight years of sobriety. He and Elisa had met at Sparky's, a preppy bar in Grosse Pointe where he sat one evening having dinner. Someone told her he was "Eric Stroh, of Stroh's Beer." She sat down next to him at the bar and asked the bartender, "Stroh's! Do you *serve* that crap?"

"I happen to make that beer," my father told her.

And they were off to the races.

All the years growing up at the mercy of his alcohol-driven mood swings came back to me in flashes of pain, and I partitioned myself off even further from the barrage of bad news.

As the weeks passed, I found I had no tolerance for the Elisa reports from other family members. The frequent calls were becoming a distraction. London was my golden chance to finally get away, to become independent, and I was determined to circumvent this landslide of family drama.

"What should we do?" Whitney asked too often, the transatlantic static engulfing his voice. "I heard she quit her job at McDonald's when she met Dad."

"Just ignore it," I advised. "I'm sure it won't last." My father had told me he would never marry again. And I simply could not picture him with anyone except my mother, certainly not with a high school peer of mine, whoever she was.

Even my mother did not take this relationship seriously. "I saw them having lunch at the country club," she called me to announce. She paused ceremoniously. "I don't think you have to worry, Frances; I saw the girl. Dad is just enjoying himself."

By this point my mother had married Lloyd Marentette, whom she'd known for forty years. The previous spring, they'd been part of a tour group cruising the Seychelles. When Lloyd had grown seriously ill with a respiratory infection, my mother nursed him back to health, sitting in his stateroom through the night and monitoring his breathing. They fell in love. Now my father could no longer drop his laundry off at my mother's house or take her out to dinner. He must have felt abandoned, all their ancient rituals finally settling into history.

Soon my MA program began, and the noise from the States receded into the background. My flat was located just around the corner from my graduate studio at Chelsea College of Arts. I worked every day, except Sundays, when the studios were closed, and bonded with the other graduate students who haunted the studios on Saturdays.

In October I audited a theory class in the undergraduate program, located just off the King's Road. I didn't recognize the instructor's name but my well-read, cynically witty studio mate, Mike, didn't miss a trick. "He's a rock star, Frances," Mike said, tossing his head up in approval. "Most important critic in London."

Students spilled into the hallway when I arrived to class on the first day. I slipped through the crowd and found that all the chairs in the seminar room were taken. Some students sat on top of a broad table pushed against the wall, and I seized the last spot, sitting Indian style with my notebook on my lap.

Standing before us was a fit man in his early forties: the famed art critic Trevor Atkins. He had cropped salt-and-pepper hair, glasses, and an ordinary face made exceptional only by the rather pained expression he wore, as if his very popularity were a source of embarrassment to him. His hands rested tentatively on the back of a chair as the masses settled themselves.

When he finally spoke, I was transfixed. He had a gift for the oblique, the ungraspable, managing to synthesize everything I had ever read or thought about art over the years, while somehow creating his own art form with the parallels he so astutely drew between art and theory.

The seminar left me feeling only more motivated; I wanted to participate meaningfully in this dialogue between art and theory. I wanted, too, to know this Trevor Atkins, to absorb his influence, and sought out conversations with him whenever possible, even arranging studio visits with him for all the MAs. Then Trevor and I arranged a theory seminar for the MA program, inviting all the top theorists in London. Suddenly I had access to people I'd only read about in the States. These luminaries came out to the pub with us after class and, later, to my dinner parties. I had finally landed exactly where I needed to be, it seemed.

The change of longitude acted as a kick-starter for my

art. I worked feverishly day and night to keep up with the intense flow of ideas, making small installations in my studio as sketches for larger pieces. I spent hours each day slicing up Dennis Cooper's *Frisk*, line by line, just as Cooper's protagonist sliced up his victims' bodies. I relabeled beer bottles with the dismembered text and sold them as art objects at a pub on the King's Road. Only much later did the piece's seemingly disparate links between the devaluation of the artist, human carnage, and my family's brewing history become clear to me.

I hardly slept. I'd wake in the middle of the night and scribble ideas inside a journal. Everything I'd ever thought or read was taking on visual shape, and I had to catch it all, as if I could outrun my terror of something essential slipping away. Come morning, I'd discard almost everything or morph certain aspects of ideas into new ones that might bear fruit.

One sleepless night the sky cracked with thunder, and rain pelted against my bedroom windows. I dropped my journal onto the floor and walked through my conservatory out onto the tiny terrace, the rain coming down hard on my head. I raised my arms up toward the pink glowing dome of London sky. I caught the rain in my hands, shallow pools forming in my palms, and splashed it over my face and neck like holy water. I stood there a long time, feeling the warm rain soak straight through my T-shirt to my skin. The muffled sound of traffic horns traveled over the buildings to merge with the rumble of thunder splitting open the clouds. I could stand there as long as I wanted, I realized; my life felt utterly my own.

. . . .

Trevor Atkins and I had been in a darkened seminar room for hours. I projected slide after slide of my work from 1990 to 1995 while he hit me with searching questions. Certain slides we'd linger on for fifteen minutes or more, discussing my ideas, my intentions. A bold rectangle of light hit the wall when the last slide had finished.

"I find it refreshing," Trevor said, "that your work is not exactly . . . 'female.' "

I wasn't sure how to take this. My breath quickened, erratic, imprecise. I hadn't been able to sleep the night before, and in anticipation of our meeting, I'd stayed up reading Trevor's exhibition catalogue essays. Compared to his brilliant lectures his writing style seemed stiff; and yet attaining his respect felt critical.

Trevor shifted in his chair. "Then again, certain pieces are clearly about how you perceive others in *relation* to yourself. Right? There's a self-consciousness to them, really, and this, I suppose, *could* be seen as 'female.' "

He was referring to my family piece, I knew, among others. "My work does tend to deal with the relativity of perception," I offered. "You know. And with point of view."

I hated the idea of my work being identified as "female"; I wanted Trevor to see me only as brilliant, apart from my sex. I yearned for this with every synapse in my brain and every cell in my body, as if my self-realization as an artist were entirely dependent on his favorable opinion of my work.

I lifted the slide tray from the projector, and Trevor

reluctantly picked up his canvas book bag. He had been in that room with me for two and a half hours, three times the allotment for a studio visit, which is what this was, technically. Much of our relationship took place in situations like these: darkened rooms with projected slides; empty theaters after assemblies or lectures had finished; hallways or corners of graduate studios where no one walked or worked nearby. We would always linger on a bit, talking about Derrida or Warhol, or my newest installation, maybe; Trevor's book-in-progress. But the distance he kept whenever others were around became confusing to me. Only when I discovered he was married did I understand. Accepting the inherent distance, I resolved to channel my interest in this man into my work itself, spending longer hours yet in the studio.

One fog-draped night, I attended the Fulbright Coalition's Christmas party, a semiformal affair in baroque quarters off Bond Street. With its arched, rooftop spires and two-story windows, the house resembled a Gothic cathedral. Inside, a butler took my coat and gestured toward a cavernous living room where, hovering above the crowd, the branches of a towering evergreen were decorated with tiny white lights and festive ornaments. Servants carried silver trays of hors d'oeuvres and glasses of red wine. I glimpsed only a few of my fellow Fulbrighters. In my patent leather boots and faux fur coat, I clearly stood out, in this room full of strangers, very conservatively dressed, very British, as "the artist."

The majority of the Fulbright scholars, all of us having come to London to pursue a year of graduate study in the humanities, had attended Harvard, Princeton, and Yale. The standard scholarship funded nine months of coursework and living expenses, but my MA program at Chelsea ran a full twelve months. To cover the difference, I had applied for a three-month extension on my grant and was still waiting to hear if the supplement had been rewarded.

Across the room, I spotted James Rutherford, director of the Fulbright Coalition. Striking, with his dark, tailored suit and perfectly combed hair, Rutherford deftly maneuvered through the crowd, chatting up distinguished-looking guests, making swift introductions along the way. The Fulbright Program was just expanding its focus from the humanities into the realm of international business, and James was overseeing this transition. At our last Fulbright meeting, he had urged the scholars to make a point of getting to know the business crowd who would be present tonight.

"Good evening, Frances," bellowed James as I moved to intercept him. "How's the art coming?"

I told him I was having a very productive year and thanked him.

"This is Bruce Lakefield," said James, introducing a bored-looking man who'd just materialized to his right with a deference that meant Bruce must be a major donor, or a potential one. "Bruce, this is our *artist*: Frances Stroh."

I'd been told during the orientation in September that artists were only very rarely granted Fulbrights and searched James's face for any sign that my additional grant might be in jeopardy, but he had already turned toward another group.

"How do you do?" asked Bruce in a clipped American accent as he offered me his hand, his expression unmoved. He had removed his sport coat and wore a white-collared shirt with a silky-sheened necktie.

"You're an American," I said, for lack of anything else to say.

Bruce had lived in London less than a year, he told me. He asked me how I was enjoying living here. This would be my first Christmas not spent in Michigan, and I suddenly felt very grown up.

"Wonderful," I said. "The Fulbright Program has made the experience of living abroad, you know, very comfortable. They treat us like royalty." I didn't tell him that when I returned from this swank Fulbright gathering, it would be to an unheated art studio at Chelsea College, where I did my work in a down coat, ski hat, and gloves.

Bruce asked me question after question about the program, and I talked it up as if my life depended on it, hoping he might tell James just how enthusiastic I'd been, how I had convinced him of the Fulbright Program's inherent value. Finally, after listening intently to my gushing, he reached into his pocket and produced a white business card.

"Frances, I'd like you to come work for me," he said with conviction, handing me the card. "You have a real talent for sales. I could *use* someone like you on my team."

I took the card. "Bruce Lakefield, CEO, Lehman Brothers International," it read. Although flattered, I laughed. Did he think I'd just ditch everything for an investment-banking job? Hell, I'd been approached by two of the best galleries in London—Lisson Gallery and Interim Art. It wasn't like I needed a job. What I *needed* was that last installment of grant money. I shook my head. "Apologies, Mr. Lakefield, but . . . I'm an artist."

He looked at me with astonishment, and then he reached out to shake my hand. "Well, you'll let me know if you change your mind."

ELISA KEYS, 1996

(by Eric Stroh)

Nearly a year later, on a soggy November night, I left my Fulham flat and made my way to the Ritz to meet my father and Elisa. The Tube was stale with unwashed commuters and the damp of the still night air. I came out of the station at New Bond Street wearing the only decent dress and jacket I had in London, feeling waiflike next to the chic coiffed mannequins lording it over every single shop's window. The Ritz stood in the distance in all its cheesy extravagance. Would she be familiar, Elisa, I wondered? "I think you'll like her," my father had said several times over the phone. "You two have a lot in common." I had no idea what to expect.

Between completing a series of installations, finishing my MA program, and signing a lease for a shared art studio with a group from Chelsea College, I'd been too busy to engage much with family matters. I'd managed to keep my promise to myself, mostly steering clear of romantic relationships, although the difficulties I'd experienced with British men had certainly aided me in this goal. From what I could tell, no one actually dated in London; they just got drunk at a pub, stumbled home, and "shagged."

The Brits' romantic side, such as it was, expressed itself through a cultural phenomenon that seemed unique to them: unrequited love. In England, everyone pined for someone—a lost love, a married love, or some other form of the impossible, and I was no exception. While at Chelsea I'd become infatuated first with Trevor and then with an artist who taught in the MA program: John Hilliard. And yet we were only friends. We often met at gallery openings and afterward would occasionally go to his house for a drink, where I'd put one of my favorite Bob Dylan albums on the record player—*The Times They Are A-Changin'*. John was the last person I knew with a cherished (and unironic) vinyl collection, all alphabetized. Substantially older than I, and with a successful international career as an artist, John seemed a role model, though, in truth, I found him at least as hip as any of my actual contemporaries. We'd sit on his Italian sofa and talk about art, films, bands, about all the people we knew who were doing noteworthy things.

And yet, as with so many men I'd met in London, John maintained a flirtatious remove that was unsettling, and I began to suspect that my unrequited feelings might simply be part of the cultural soup in which I found myself. I was learning to speak a new language—the language of calibrated distance—with the hope that our carefully monitored exchanges would somehow lead to more closeness.

In spite of how lonely this was, my life would undoubtedly

have appeared, to the outside observer, rather glamorous—full of parties, dinners, gallery openings, and interesting friends. Life moved quickly in London, and with all the distraction, I'd found one could easily spend a year or two—or five—and still be alone.

And so a part of me felt relieved that my father wasn't alone. Though I spoke with him less frequently since Elisa had entered the picture, I realized his preoccupation was a recognizable sign of happiness. My grandmother, whom I badly missed, was the only family member with whom I spoke regularly. The rest had gradually stopped calling to report on my father's comings and goings. But when my father called in October to say he was bringing Elisa all the way to London just to meet me, I knew things must be serious.

Stepping into the hotel, I recognized Elisa immediately. She and my father were sitting on a round silk sofa beneath an enormous crystal chandelier at the center of the lobby. "Christ," I said to myself. "It's Eat the Rich." In high school, I had never known her actual name. I remembered standing outside a heavy glass door by the school's parking lot with the smokers, cracking jokes between drags and getting laughs from the crowd gathered there, Elisa among them. She wore the same clothes every day—a moth-eaten army fatigue jacket with the words *Eat the Rich* embroidered on the back in fiery sweeps of red and blue. Her suede lace-up moccasins just reached the knees you could see exposed through shredded holes in her jeans. Her deep voice and towering height would have deemed her an important member of the crowd, though I rarely heard her speak. She seemed always to be lurking in the hallways or striding angrily across the school's lawn as if chased by tiny devils with blazing pitchforks. And that emblazoned message . . .

My father waved me over, and I felt my legs propel me forward with involuntary momentum. This girl's style had certainly changed, I thought to myself, taking in her fitted tweed suit, nude stockings, and brown tasseled loafers, a single strand of pearls around her rather thick neck. My lips strained into a smile. I watched her eyes. They were, I saw, the eyes of a cornered animal. Was she hoping I wouldn't remember her? Though both of us were nervous, clearly, I felt worse for Elisa somehow.

My father stood up, smiling, and leaned toward me for a hug, his pipe clenched between his teeth. "Hello, Franny," he said with determined cheer. "I like your *hair*!"

As Elisa stood up and shook my hand, I did my best to put her at ease with friendly chatter, and I continued with this as we made our way toward the dining room.

"Quite a town, London," she said, surprising me with her ability to make small talk. "How'd you like living here?" Her voice was that same mannish deep tenor I remembered from a decade or so earlier. Her clothes, really, were the only remarkable change, and the fact she had aged the way drinkers do—red face, lined eyes, a little big around the middle. "Don't much like big cities myself."

"Really?" I said. "*I* love it."

"But that's great," she grinned. "What do you do here, anyway?"

Hadn't my father told her? "I'm an installation artist," I said.

"Really! Sounds like fun."

I cringed at what sounded like a somewhat condescending tone. Did she even have any idea what an installation artist was? My father stopped and fumbled in his breast pocket for a lighter. Elisa reached into her handbag and smoothly produced a purple Bic to relight his pipe. "Goddamn limey tobacco," my father mumbled as we came into the fancy dining room. "Whole *country's* damp."

"We spent the day at Dunhill's," Elisa told me with a proprietary indulgence. "He's trying out a new tobacco."

"Just like old times," I said. "Dad and I used to spend the whole day at Dunhill sometimes when I was a child." Unsure how to continue, I looked over to my father, just then busy with the maître d'. His spending the day with Elisa at one of our old haunts felt like a betrayal, and I sensed this would be the first of many times I'd feel this way.

Back in high school, I remembered hearing once that Elisa had been in a car wreck with her mother and that her mother had died, a fact that had elicited my respect and some deference in matters of suffering, though I couldn't have imagined then that our separate longings for a parent would one day collide.

Soon we were seated at a round table at the center of the dining room. Our French waiter was one of the most handsome men I'd ever seen, and his flirtatious attention helped distract me. Still, I caught myself realizing, as the drinks were served, that Elisa, who'd been something of a wallflower back in high school, seemed determined to assert her new authority with assurance.

"So, do you get paid for that?" she said. "The installation stuff?"

My father sat smiling complacently as he nursed his vodka and ginger ale—a drink I hadn't seen in his hand in ten years—seemingly happy that Elisa and I were "getting along," probably telling himself that all his worries about the introduction had been for naught. I wasn't sure which I found more disturbing—the fact my father was drinking again or that he'd brought that girl from the smoking exit all the way to London. The two events seemed fundamentally connected.

"In a sense, yes, I *do* get paid," I told Elisa. "I came here on a grant, see." I sipped my wine and hoped she wouldn't ask me what, precisely, "installation" meant.

She took a mini baguette from her plate and tore off a bite with her teeth. "So, what *is* 'installation,' anyway?"

"Can they make me a hamburger?" I heard my father asking our waiter as he gestured toward the kitchen. "*Well done,* please."

Elisa poured herself a glass of wine before the waiter had time to come around to our side of the table. The waiter took the bottle from her hand and finished pouring. She smiled at my father, who wheezed through the smoke as he lit his cigarette. He looked tired.

"Installation art," I told her, "well, it defies boundaries or strict definition, really."

"So it could be . . . anything."

"I suppose it *could* . . ."

"Frances is a damn good photographer," my father said. "I wish you'd get the old camera out more often."

I smiled and lifted my glass of wine. "Cheers," I said, and we all toasted, to what exactly I was not sure.

My father and Elisa had come over on the *Queen Elizabeth 2* and would return on the Concorde. My father was rolling out the red carpet for Eat the Rich, and all I could do was drink my wine and act as if this were all perfectly normal. Elisa and I shared a bottle of cabernet while my father sipped his vodka.

After dinner, my father retired to bed, and Elisa and I went out to a nearby nightclub for a drink—her idea. My father chuckled, waving us off through the closing elevator door. "Don't make it too late, girls."

The club was very West London posh—low lit, ritzy, and full of sexy, well-dressed people out to be seen. "Cool place," Elisa said, not without a certain discomfort in her voice.

Installing ourselves at the long lacquered bar, trying hard to be heard over the Portishead blaring in surreal, hypnotic waves, we ordered scotches on the rocks. The bartender scrutinized us as he poured. Was he trying to make out our relationship? Elisa took a long draw from her drink, emptying half the glass. We talked about people we'd both known in high school and what had become of them, although in truth there wasn't a lot of overlap, beyond the smoking exit.

"You knew Caitlin Jaspers," asked Elisa. "Right?"

The bar was filling up, and I knew it had to be around

midnight. "Of course." My mind flashed to the pale, impossibly beautiful girl in Hobey's backseat, her thick black hair blowing around as we headed to the Uniroyal plant.

"She's really been through hell," said Elisa. A while back, she told me, Caitlin had gotten out of the car one night at the side of the freeway after a fight with her boyfriend, then was badly beaten by someone who came out of nowhere. By the time her boyfriend came back to get her she was being loaded into an ambulance. Her face had to be totally reconstructed.

I found the story so upsetting I felt physically ill. "Oh, my *God*," was all I could say.

"Yeah, it was really bad," continued Elisa. "You might not recognize her now. But she got married, you know . . . now she's teaching photography or something at a high school in Maine."

I found even this happy ending depressing. I'd imagined so much more for Caitlin—a glitzy career as a *Rolling Stone* magazine staff photographer, rock-star boyfriends she'd stub out like cigarettes as soon as she grew bored, a wardrobe that would put Kate Moss to shame. "*Damn*," I said. "I'm just . . . floored, I guess. Caitlin deserved . . . so much better."

"I know." Elisa nodded toward the back of the club. "Hold on. I'll be right back."

I had thought about Caitlin often since high school. I'd last seen her ten years before, at a Beastie Boys concert at St. Andrew's Hall in Detroit. She was photographing the band for a magazine spread. The glamorous life I'd envisaged seemed well under way. And then, apparently, it had all come apart. Or had she just realized she didn't need all the lights? Didn't I myself have some of the same doubts?

As flamboyant a partygoer as I could be, the truth was, I liked the quiet life of making art, talking about art, and enjoying close friends at dinner parties—the kind of life I'd been having in London for the most part. Those evenings I spent out among the "art stars" at private views and parties, they weren't the most memorable, the ones I truly cherished.

Elisa weaved through the dancers on her way back from the restroom, making a goofy show of using her handbag as a front fender to ward off collisions.

"I'm still speechless," I said as she sat down.

"Yeah." She studied me, her eyes widening in the dim light. "Not to change the subject or anything, but . . ."

"Go ahead," I told her.

"I've been feeling really sorry for your dad. He's a good man, but . . . you know, he never got any *love*." Her voice cracked on the word *love*.

"You're right . . ." I said guardedly. I wasn't sure where she was going with this, but I had the distinct feeling I was being rather purposefully led. "He got shortchanged, I guess, as a kid."

"See, so did I," she said, downing the rest of her drink. "I know what it means to basically be an orphan. I'm going to make up all those lost years to your dad, though, you know? Give him everything he never got. You'll see."

My stomach tightened into a thorny little ball. She was going to parent *him*? As much as I wanted to believe this, the age difference between them had already spelled things out pretty clearly: Elisa would be the pampered child. My father had already told me they slept in separate bedrooms and that he gave her a large monthly allowance.

Besides, I had my doubts that anyone could make up the losses of childhood with another person's affection. I'd been my own lab rat, and the data was not encouraging. "Do you really think it works that way?"

Elisa nodded slowly, considering her next move. "He's like a lost little kid. He *needs* someone."

"He should have been a photographer," I said.

"It's funny," she said, resting her elbows on the bar. "That's exactly what he always says about you . . ."

Gorgeous people swayed all around us to the heroin beat of Massive Attack. A couple fell against the bar in a cinematic embrace, one of her long earrings coming unhooked as she swatted flirtatiously at his drunk, groping hands. Elisa lit a cigarette and I saw her, eleven years before, huddled in the corner of the smoking exit, laughing at my jokes but perhaps afraid to speak, the snow blowing around us in subzero gusts.

***NON-SPECIFIC LOCATION #1*, 1996**

(by Frances Stroh)

My shared art studio in Southeast London had a large window facing a block of boarded-up storefronts and crumbling prewar brick warehouses. I often paused on the windowsill in the afternoon to absorb some heat from the old radiator beneath it, looking out. There were no trees anywhere, even on the vacant lots. The brown grime covered every surface—of ancient coal dust, or an eternity of exhaust fumes. The very people milling about outside seemed to have absorbed it into their drab clothing, their worn-out shoes.

I was working on a new idea: rudimentary remakes of famous film scenes. So far I had made just one—the scene in *Apocalypse Now* where the helicopter lands on the beach and blows everyone's stuff away. I had reconstructed the beach on a large rectangular table using sand and cocktail umbrellas, then simulated the flight of a helicopter by coming down onto the beach with the video camera and a hair dryer blowing from behind it. The effect, with the addition of an audio track of a helicopter in flight, was magnificently absurd, especially as all the umbrellas blew away one by one.

I hadn't been able to think of any other movie scenes I wanted to re-create. We had an open studio show coming up in early December—only one month away—and were committed to exhibiting at least one installation. All week, while I'd been busy showing my father and Elisa around London, my studio mates had been hard at work on their pieces. My father was so jazzed on Elisa he hadn't even remembered to suggest a studio visit to see my work. Then my flat was robbed, tying me up for days at the American embassy, where I'd gone to replace my stolen passport. The thief had also taken the Nikon camera and lenses my father had given me in high school, all my personal photos, and some cheap jewelry.

"So you were *burgled*, were you?" the police officer exclaimed when I'd called to report the break-in. But they never came out to investigate. George Orwell, I figured, had the British attitude down: property itself was theft, which was clearly why I'd been relieved of mine. Case closed.

By my father and Elisa's last day in London, I was eager for them to board their Concorde if only so I could get my life back. It had been two weeks since I was in the studio. On the very last day, while Elisa had a massage at the hotel, my father and I went to see the Degas exhibit at the National Gallery. Standing before the hundredth painting of ballerinas, I pictured lining them all up on the beach and gunning them down. I knew I could get the soundtrack of a machine gun somewhere, and music box figurines of ballerinas were plentiful. Perhaps my idea needed some evolving. I wouldn't limit my imagination to scenes from movies. Perhaps anything was fair game.

My father grunted at the painting and wandered off. I trailed behind him. Passing a tall ballerina sculpture, he looked shrunken beside her proud, lithe stance on the pedestal, his shoulders caved forward. He was getting old. I wanted to reach out and hug him through his khaki raincoat. But I didn't.

The day before, we'd been at Harrods, where my father had bought Elisa a leather backpack costing £400. Afterward, while they looked on uncomfortably, I tried on a pair of earrings that cost £100. My father seemed torn, with Elisa watching, between pulling out his credit card and letting me buy them myself. It occurred to me, then, that any friendship that might form between Elisa and me would be contrived only for my father's sake.

As I admired the gold earrings in the mirror, Elisa ruffled the tissue paper in her enormous Harrods shopping bag. The tension was stifling. My father wheezed through his cigar smoke, wondering out loud where we should have lunch. Finally, I pulled out my own credit card and handed it to the sales clerk. My Fulbright extension had just ended, and I could hardly afford the earrings, but I wanted to show my father that I could take care of myself, even as I was beginning to wonder if it was true.

Later, when Elisa was seated at the hotel bar downing yet another drink, my father led me into the lobby and gave me an envelope containing £400. "Happy birthday," he said. "Maybe get yourself a new camera." It was an apology, and

I felt grateful for it, though the money was, I knew, a booby prize.

I spent every penny of the gift on black cabs to and from gallery private views, instead of taking the Tube, a luxury I rarely allowed myself. The money lasted just two weeks, until my thirtieth birthday had passed. I knew I was being reckless with my father's money, reckless the way he was, and this somehow calmed me.

"Elisa and I are engaged," my father announced over the phone, just one week after their Concorde flight had touched down in New York.

I sat down on my futon sofa and studied the chips in my thrift store coffee table. "That's great, Dad."

"And look, you don't have to worry. We'll be signing a prenup. Bill Penner's drawing one up."

"I'm not worried about *that*." I hadn't even considered it, in fact, but felt reassured just the same, knowing that Bill Penner, the family lawyer, would handle the matter well.

"I hope you'll come to the wedding, in the spring."

"Of course," I said.

"And I'm inviting you kids to join us out in Jackson Hole after Christmas. My treat."

"Wow, thanks." We hadn't been on a family vacation since I was a teenager. But Charlie would be excluded from the trip, I knew, and I wasn't sure I was ready for the family politics, particularly now. "Can I let you know as we get closer?"

"Of course."

I felt surprisingly unburdened as I hung up the phone. With this frightful thing actually happening, there was at least one less catastrophe to fend off. I even felt happy for my father, remembering how Elisa had smiled at him at dinner, lit his pipe for him, laughed at his awful barroom jokes; she had a few of her own to share, too. Maybe it made a kind of sense, this odd May–December match.

The birds on the brick wall outside my conservatory chirped in the smog. A fake-sounding British siren bellowed down Wandsworth Bridge Road. Soon I'd go over to the Indian liquor store to buy my bottled water, and on my way home I'd stop to chat with the furniture makers who rented the storefront below me. What did it matter what went on in Michigan?

With just two weeks left to conceive and finish an installation, I still hadn't decided what I'd present in the studio show. Usually I would spend months on a new piece. Knowing now that this effort would be rushed, I was reluctant to invite Trevor Atkins or any of the art dealers.

Trevor, who was moonlighting as a guest curator, had put some of my British friends into a show at the Hayward Gallery—but not me. When I'd run into him at a gallery private view in early November and asked him about his decision, he'd said, "It's a survey of British artists, Frances. Nothing personal." But it did feel personal, particularly after Trevor's enthusiasm during my final MA exhibition at Chelsea back in September. The buzz from the visiting art dealers and critics had also been encouraging; Lisson Gallery and Interim Art had made special studio visits to preview my work. A mere two months later, Trevor was putting safe—and mostly male—British "art stars" into the Hayward Gallery show, and the art dealers, too, were radio silent.

I began to wonder if I was too much of an outlier as an American artist. When a well-known British artist offhandedly suggested I keep my contacts active in the States, "just in case," my suspicions solidified, and I began to believe that coming to London had been a grave misstep in my career.

I should move to New York, I decided. Then again, the thought of establishing myself in yet another big city was exhausting. To say nothing of the expense. With my Fulbright income gone, I needed a teaching job in a college art department. The small income I received from our real estate trust in Detroit, while helpful, was not nearly enough to live on. I began to have fantasies of escaping the drudgery of life in London to some whitewashed village in Greece. There, I could build up reserves—both emotional and financial—before my next move.

With my momentum in London all but exhausted, I often forwent the hour-long Tube ride to the studio to work on a piece for the upcoming show. Instead, I became involved with projects around the house, like applying for studio programs in New York, or vacuuming. Or polishing camera lenses—the used ones I'd bought to replace the stolen lenses.

At last I arrived at an idea for the studio show: a life-size video projection of a poker game. It was part of a series of "happenings" that I filmed from the ceilings of rooms and then projected, often back into the very spaces where the events had been recorded.

Having spent the afternoon at the studio setting up for the shoot, I pulled a rectangular table into the center of the room, climbed on top, and attached my video camera to the ceiling. Then I spent an hour adjusting the angle of the camera, observing the image of the tabletop on a monitor, lining the edges up with the frame.

With the show only a few days away, an oppressive mood had settled over the studio. Dispirited, Mike and Ole tinkered for hours with a broken video projector. Tanja came over from across the hall to use our bathroom, but didn't stop to talk. Gary sat at his desk, ignoring everyone.

My poker-playing friends were due at any moment, but then the phone rang and I went over to answer.

It was my brother Bobby. "So, dad eloped with that . . . thing."

I didn't know whether to laugh at this, or cry. I leaned into the wide windowsill, observing the desolation of the street outside, dead grass studding the empty lot beyond like patches of mold on old bread. Tiny flecks of rain blew sideways through the air.

"Where?" I said.

"At Gruhn's—that redneck guitar store in Nashville."

I knew it well, the store. My father spent tens of thousands there each year buying vintage Martin guitars and selling others back for pennies on the dollar. "I thought he was going to invite us to a wedding or something."

"Well, he was with Elisa and Ginger Boatwright and they all got drunk and threw an impromptu wedding. No prenup. Nothing. Bingo, right? Mom is fucking *pissed*."

Ginger was a buxom and soulful bluegrass singer, a friend whose career my father had probably subsidized. She herself had wanted to marry him at one point, but he must have been holding out for a young thing like Elisa.

Hanging up the phone, I felt as if a truckload of wet concrete had just been dumped on top of me. I wondered how I'd make it through the shoot.

I walked back to the set and adjusted some floodlights around the poker table. Neither Mike nor Gary looked up from his work. Ole watched me silently for a moment, then returned to the broken projector. The old radiators began to kick out some heat, and the room grew warm with the comforting smell of scorched dust.

On the day of the studio show, I stood at the center of the exhibition room, observing my piece for the first time; just a color projection of poker chips and cards, with hands manipulating them on a tabletop. No sound.

With its banal subject matter and lack of implicit critique, the piece came across as . . . nothing more than itself. Rather than enlarging the experience, the all-seeing eye of the camera had reduced it, and I was relieved no one mentioned it to me either during or after the private view, which attracted over a hundred friends and art-world types.

In the disappointment of the show, I couldn't help but feel the tide had permanently turned for me in London. If the proverbial club sandwich had ever been there, it certainly wasn't anymore; and I no longer felt deserving of success, anyway.

My flat was a warm, safe cave in which to hibernate through another London winter, and I stayed home for days, subsisting, like my mother, on cream of tomato soup from cans, and sleeping too much. The answering machine picked up my messages with the volume turned all the way down. My clothes sat in piles on the floor.

It seemed I was losing everything all at once: my father, my family, what I'd hoped to be my career. So many blows in so short a time. And that legacy of failure I thought I'd left behind in the States? It had not only caught up—it had overtaken me. I wasn't the favored one anymore, not anywhere, and it felt like a kind of death, as shameful as it seemed to even think this way.

One day, when the phone rang, I decided to answer it.

"Frances," said my father. "Will you be coming home for Christmas?"

"I don't know, Dad." Sudden tears burned my cheeks raw. "I'm just . . . you know, really, really busy here."

"What's *wrong*?" he said.

Nothing, I told him, and we talked instead about the weather and the beer business, avoiding entirely the matter of his marriage. Elisa, I figured, would be right there in the room beside him, downing a beer, or firing up a smoke. I had to choose my topics carefully. Now I would be the one hovering in the shadows, afraid to speak. The absurdity of the thought made me smile for the first time in days.

. . . .

I was stretched out on John Hilliard's sofa in furry leopard pants. John, who lately had taken on something of a fatherly role, sat in the chair to my left, talking over my future with me, red wine and crackers laid out on the table between us. All around us, printed on enormous canvases, were John's famous photographs, resting against the tables and walls. I adored my friend's house, formerly the vicarage of an Anglican church, with its Gothic arches and minimalist decor.

"London can really work for you, if you give it some time." He sipped his glass of cabernet. "Make three more of the inkjet prints, why don't you, and invite the galleries over to your studio."

As part of my final exhibition at Chelsea, I'd made a nine-by-nine-foot color print—a partial aerial shot of an unidentifiable town in the North of England—enlarged and sumptuously printed on vinyl, then mounted on stretcher bars in the manner of a painting, much like John's work. He had assisted me in stretching the "canvas," an all-day undertaking, and the print had been the focus of much attention during the exhibit. Now it was rolled up in a closet at my studio, collecting dust.

I knew John believed in me the way I'd once believed in myself, but it felt too late somehow; I just couldn't rally any more optimism around the prints, or, in fact, much of anything. "Those prints cost eight hundred pounds apiece to make," I said dismissively. There was no way I could afford three more.

Instead, I was working on another video installation, panning my camera across still images of landscapes I'd found in books. Edited together, the effect was that of the view from a train window, a dizzying assemblage of nameless places passing by—only I was finding that gaining perspective on false constructs was a far simpler feat in art than in life itself. In life, the false constructs themselves tended to take over.

"Anyway, I'm going home for Christmas," I told John. A kind of masochistic curiosity had seized me, and I'd booked an airline ticket. "Maybe I'll make those prints after I get back," I added, though I knew I wouldn't. I could feel something closing off inside of me, the drive to make pieces waning, and well before I'd gotten the recognition I'd so badly wanted. Being in favor had always been my mode of survival and, later, the brittle foundation on which I'd built my creative life, my independence. Now, though, all that had changed. I'd need to find something else to lean on.

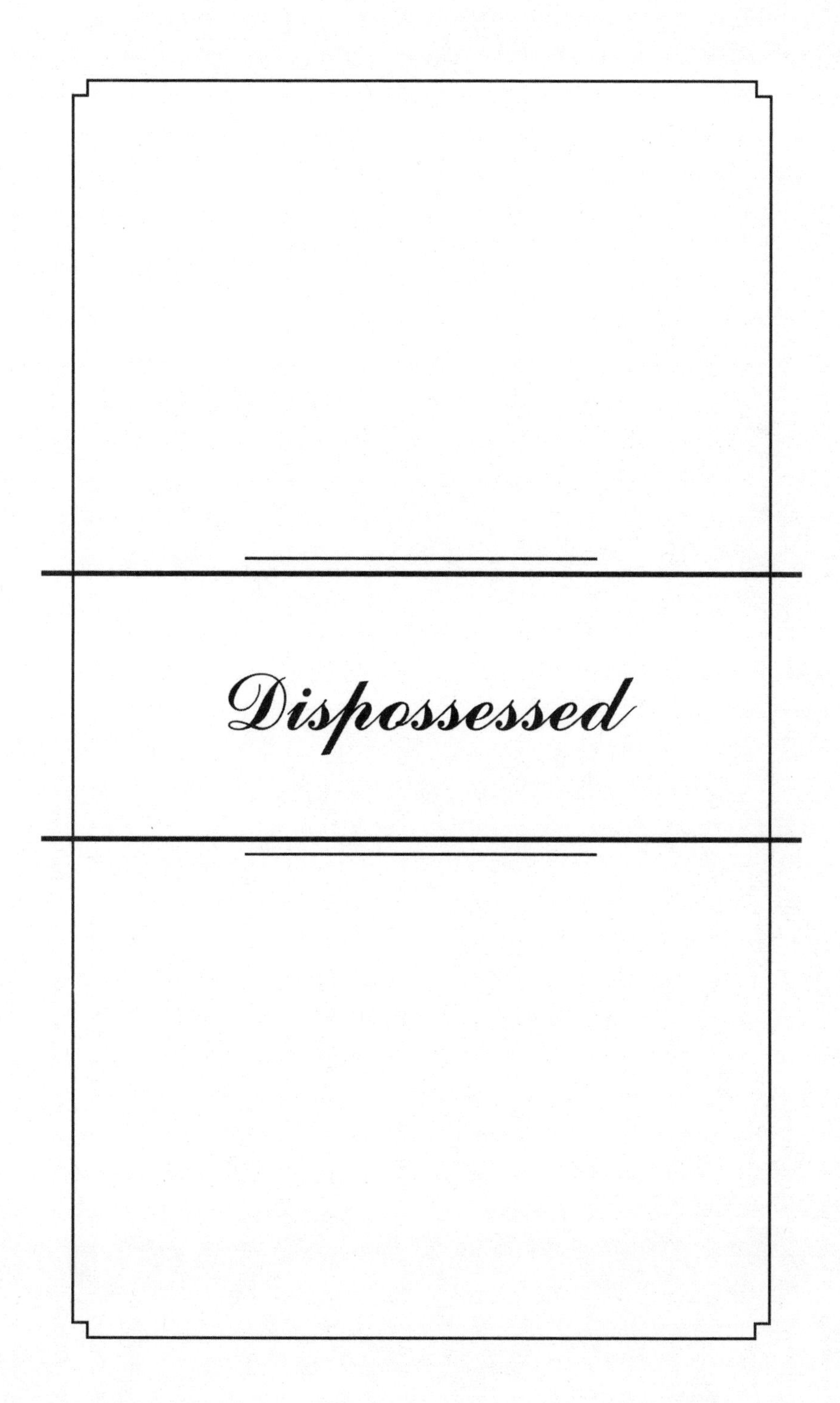

Dispossessed

LONDON, 1996

(by Tanja Merz)

Grosse Pointe, 1996

My first night back in Michigan the sky dumped a magical foot of snow, and the morning cast a pale-blue light across my room. Hearing my grandmother's voice downstairs in the kitchen, and catching the scent of coffee, I sank back into the pillow supremely comforted.

I'd been woken by the shrill voice of a neighbor dropping off a present at my mother's door. "It's supposed to snow *three feet* before Christmas," the voice had caterwauled. "*Three feet!*"

And then, "Gailie, close the door," my grandmother had shouted from the kitchen, silencing the voice. "*I'm cold!*"

It was my grandmother I had missed most of all during my year and a half abroad. Where my parents had fallen short when we were young, my grandmother had always been ready to step in, nurturing us, sometimes even spoiling us.

Smiling at her high-pitched laughter down in the kitchen, I quickly dressed, brushed my hair, and went down to greet her, taking in the French perfume, even from the top of the stairs, that always marked her—Joy by Jean Patou.

Passing through the dining room, I could see her sitting at the kitchen table in the next room, regal in her puffed mink hat and mink coat, which she had draped over herself like a blanket. She always complained of the cold in my mother's house. Her wispy white hair had been styled, perhaps just the day before, at the hairdresser she liked here. At eighty-nine years old, her still-beautiful face was free of wrinkles, thanks to her daily masks of drugstore cold cream.

"*There's* Frances," said my mother with a big smile, as if introducing me on one of those festive afternoon talk shows.

"Hi, Granny," I said excitedly, bending down to kiss her.

My grandmother looked up, her soft brown eyes bright and expectant, but her smile quickly faded. "What have you *done* with your *hair*?" she cried. She turned back to my mother for an explanation.

My mother said nothing.

"It's short, Granny. That's all." I knew she was prone to outbursts of scathing criticism, but thus far I had avoided being a target myself. I turned on the kettle for tea, trying to ignore her condemning gaze.

"But it's just . . . *awful*," my grandmother said. "Look at you! You used to be so beautiful, Franny, and now . . . now you're just a plain Jane!"

The last time she'd seen me I'd been a long-haired golden blonde, it was true, but in London I had wanted to shed my past, to create myself anew, my hair getting shorter and blonder with each salon visit. I poured the hot water into the mug and dropped in a tea bag. "Granny, come on, it's just that you haven't seen me in a while. I've *changed*."

My grandmother brightened. "Well, have you met anyone nice over there?"

I knew what "nice" meant. Marriageable. At thirty, I'd been over the hill in her book for a good five years now. "I've met a *lot* of nice people," I said with a smile, dodging the question. I certainly wasn't going to launch into a discourse on the British sensibility of unrequited love.

"And what about the Fulbright crowd?"

"I'm not going out with any of them, if that's what you mean."

She turned to my mother. "I just don't understand this at all," my grandmother said. "She had the world by the tail!"

"Take it easy, Mother," said my mother. "Frances is doing all right."

"All right?" I protested.

"What kind of art does she do, anyway?" asked my grandmother.

"I don't know, Mother. Something with video."

My own mother didn't know how to describe my art, but I couldn't blame her—I had trouble putting it into layman's terms myself. I left the room, carrying my steaming mug of tea with me. "Granny, look, I don't expect you to understand my choices," I said from the doorway. "But I hope you'll at least try to respect them. Okay?"

"I *want* to understand you," she said. "It's just I don't!"

I went back upstairs to my bedroom and crawled under the covers. Flurries of snow blew outside the frosted windows.

My mother had moved to this more modest house when I'd turned twenty-two, and noise traveled. I could hear them continuing the debate downstairs, my grandmother irate, my mother doing her best to pacify her.

My grandmother had always been my greatest fan, spoiling me in a way my mother never had. She was the one who provided the piles of fresh fruit, new winter coats, the trips to the toy store—indulgences my mother generally discouraged. My grandmother's love, it seemed, never wavered. Once I'd started college, she took me on regular shopping trips in New York, even setting me up with a personal shopper at Saks Fifth Avenue. We'd go from one department to the next, piling clothes into the personal shopper's arms. I could point to any garment or accessory, and she would carry it back to a private dressing room, where all the things would be waiting when I came in. That same night, I'd wear one of my new outfits downtown to have drinks at the Odeon, check out a band at CBGB, and head to an artist's loft party in the Bowery with a gang of friends. The following day, I'd meet my grandmother uptown at the Colony Club for lunch wearing one of the more conservative dresses she'd bought me. I could surf both worlds comfortably, but lower Manhattan, that was my thing.

"Promise me you'll never take the subway," my grandmother would say after lunch at the Colony Club, handing me an envelope full of twenties and hailing me a cab on Park Avenue. "I *wish* you wouldn't always insist on staying *downtown*."

My grandmother stubbornly held out hope that I'd fall

into the life she imagined for me: married at twenty-four, say, to a good Upper East Sider, summering in Nantucket or the Vineyard, holding whatever club memberships, maybe even becoming the kind of killer bridge player she was. Now I understood that all the pampering, all the exclusive vacations and beautiful clothing, had been an active attempt to secure this future for me. Clearly, it had not come to pass. In her eyes, then, I had failed. I was single, an artist; I had opted out of the society that was my birthright. Had I become a painter, at least, perhaps the blow might have softened somewhat. A painter was the kind of artist people understood. Renoir, Monet, Picasso. You know. "Frances is the next great painter," they could have said at cocktail parties.

"You aren't so worldly, Frances," my mother sometimes told me, meaning, I think, that I didn't steer my opportunities to any particular outcome; I was a bit of an idealist. I'd always taken this as a compliment, though I was fairly certain it hadn't been meant that way.

Whitney and Bobby arrived in the afternoon, Whitney from Palm Beach, his L. L. Bean duffel bag freshly dusted with Michigan snow; while Bobby was just coming from across town. Bobby had recently become a Grosse Pointer again, having moved from Dallas to work at the brewery headquarters. I'd primped in advance of their arrival, now morbidly self-conscious about my appearance, but also excited to see the men in the family, to whom the women's attention would now, thankfully, turn.

"You look every bit the artiste," said Bobby when he saw me. Alright—that was more like it.

"Frances, looking good!" said Whitney. "Step outside for a smoke?"

We huddled by the garage door in our sweaters, minimally shielded from the gusting snow. A station wagon crawled along the street and skidded at the stop sign.

"*This* Christmas is gonna be a fucking joy ride," Whitney said with a wicked smile, dragging on his Camel.

Because of our father and Elisa's elopement, he meant. We were all going out to Jackson together—a trip my father had clearly planned to get us acquainted with Elisa. We were leaving the day after Christmas.

First, though, we were scheduled to see my father and Elisa, right here in Grosse Pointe, the following day. "Change is in the air," I told my brother.

"No kidding." Whitney tossed his cigarette into a snowdrift and rubbed his hands together. "As if it weren't bad enough, Mom moving to this grim little house."

Whitney and I went inside to find Bobby and my mother talking at the kitchen table. Whitney's loafers squeaked on the kitchen's plasticized tile floor; he smirked in my direction. I knew he missed the old house—its proximity to the forest where we played as children, our cousins Pierre and Freddy nearby, the familiarity of every curve on the road when he sped his car home after a high school party. I understood. I missed those things, too. The old house had been part of the old life, the life we'd lost, bit by bit, after the divorce.

"Too many memories in this place," my mother had said just before she sold it.

At the new house, a 1950s brick box, we might see a neighbor doing gardening work, inflating a baby pool, or firing up a grill, activities we'd rarely seen on Provencal Road, where gardeners, cooks, and hired pool cleaners were the norm.

But my mother loved that her new neighbors kept an eye on one another's houses, talked to each other over fences, and dropped off tins of cookies at Christmastime. Perhaps she felt less isolated here. It agreed with her. And since marrying Lloyd, who was now in Oregon visiting his son, she seemed to have dropped ten years from her age. She looked slimmer and happier, though her chronic insomnia over the years had left dark circles under her eyes.

Whitney went upstairs to unpack. After his shower, the scent of Bay Rum aftershave floated in the hallway. My grandmother put extra wool blankets at the ends of our beds and turned up the heat.

Charlie, my mother informed us, had gone to Mexico for the holiday. "It's for the best," she said flatly, her eyes trained on the pot of Campbell's soup she was warming at the stove.

My mother had arranged a meeting for us downtown for the following day with Bill Penner. "You need to get the facts," she said. She meant about the financial implications of my father's elopement. Her thirty-year settlement agreement with my father would not be disrupted, but my brothers and I had no such binding agreements. "You all have to face this." One might have thought, from her tone, that we were headed for the guillotine.

We dressed for the occasion, I in a black-and-white checked double-breasted jacket with black pants and heeled boots; Bobby and Whitney, sporting blazers, ties, and gray flannels. We met in the trust department of our family bank in downtown Detroit, an entity designed to keep the family-trust management under our own roof. Nearly all the Stroh Brewery Company shares were held in multiple generation-skipping trusts for the protection of the company and the family—at least that's how my grandfather, etc. planned it. These trusts kept the brewery assets intact so that subsequent generations of Strohs might continue to own and manage the business while also paying out quarterly distributions to the family shareholders and their spouses. Within our nuclear family, my father was the only shareholder.

Bill Penner, an attorney with the big Detroit firm Butzel Long, agonized over a pile of documents at the head of the boardroom table as we filed in, his skin stretched too tightly over his clearly exhausted face. He looked up with reddened eyes. "Thanks for coming down," he said in a tone that could only be called funereal. "I wish it were under better circumstances."

We all shook his hand and sat down. Bill had grown gaunt with age. Whitney had once remarked that he looked as if he "summered on Three Mile Island," and I couldn't help but laugh. Now I felt badly; this man seemed so genuinely concerned about the situation we found ourselves in, so determined to do whatever he could to help us.

"I've been reviewing the trusts that benefit your father," he said soberly as he flipped through a thick document. "And, well . . . Afraid I don't have very good news."

The air left the room. Bobby wiped his brow and gave me a long-suffering look that said, *Here we go . . . again.*

"As you know," Bill continued. "Your father and Elisa did not sign a prenuptial agreement before they wed. That in itself is troubling. The other piece of unfortunate news, though, is this: two of the four generation-skipping trusts that benefit your father give him a power of appointment over the income, meaning that he can—or even, from a legal standpoint, may *have* to—appoint that income to Elisa after his death."

I felt the raw taste of fear at the back of my throat.

Whitney leaned back on the rear legs of his chair, his hands white-knuckling the edge of the mahogany table.

Bobby looked up from his notepad. "Let me get this straight, Bill. Our father may have to leave Elisa his trust assets? I don't understand—that money was made by his father and *grand*father."

Bill nodded. "That's correct. And just to be clear, he may leave only the income from the trusts' assets, not the assets themselves. Those, of course, will pass to your children."

We had no children. And even if we had, Elisa would receive the income for the entire span of her lifetime. My brothers and I would be passed over entirely, lending the term "generation-skipping trust" an unpalatable new twist. Until now, I had never given the trusts much thought. I was young, and artists, well, they were always just scraping by. And so I had learned to romanticize my bohemianism, if only to cope with a lifestyle that afforded few of the luxuries I'd enjoyed in the past. In fact, until very recently, I had been perfectly happy on my chosen path, self-abnegation included. But now

I realized that the Stroh trusts—psychologically certainly—had always, in truth, served as something of a safety net for me. I'd never had to cope with serious uncertainty. All that had changed now. And not just because of my father's impetuous behavior.

The vast majority of the Stroh Brewery's value had virtually disintegrated while I'd been holed up in my art studio in London, wrestling with esoteric issues of point of view in my video-installation pieces. For several years, every major brewer in the United States had been undercutting its competition with price reductions, and Stroh had been forced to follow suit; with no margins and no cash, our business was tanking. In a desperate last attempt to stay afloat—and with the help of another massive loan—the Stroh Brewery acquired the G. Heileman Brewing Company, a brewery in even direr straits than our own. The marriage was a poor one, and our combined sales continued to drop. Yet the Stroh board went on paying the family shareholders the large income to which everyone was accustomed. Indeed, my father's lifestyle seemed to grow more lavish by the minute, with grander houses, fancier boats, and showier cars. His denial was contagious; even Bobby, Whitney, and I assumed our businesses must be faring better than reported at the annual family business meetings. For years, we'd all flown cross-country to attend these meetings—a requirement of all shareholders, present and future—under the assumption that the trusts would be there and that we, as future shareholders, were ultimately responsible for our stake in the company. Now it was clear we'd never had any control over the company's destiny, let alone the trusts.

"Jesus, this just keeps getting worse," said Bobby as he read the bullet points Bill had placed in front of us.

"Each of you will have to seriously consider what you will do for income," Bill said, sitting back down. "Historically, the brewery has supported the family, but you would not be well advised to consider these trusts as . . . a significant source of future income."

It was as if I'd come home to a seared patch of ground where my house had stood. Nor was the problem simply the loss of future income. Suddenly, the notion of being an artist seemed frivolous and misguided; I'd have to find something to apply myself to that guaranteed a decent living. Hell, I didn't even have medical insurance. On some level, certainly, my mother had strived to prepare us for this all our lives, but being actively disinherited, well—this was a rather different thing from having money and pretending as if one didn't. The life I had imagined for myself—becoming a successful artist, owning my own apartment, perhaps even collecting the work of other emerging artists—suddenly it all felt well out of reach.

Reading the fear in my brothers' faces, I realized they, too, must be letting go of certain hopes about their futures. For Whitney, it was an Upper East Side apartment, Augusts on Long Island, and the ability to leave behind that monthly spike in anxiety triggered by opening his American Express bill. Just the night before, when my mother asked him about his job raising money for a friend's company, he'd quipped that he'd "rather live Unabomber style in the woods of Montana" before he did any "bootlicking." Now he might just have to.

Bobby had hoped one day to be able to leave his job at the brewery and run a restaurant in the Caribbean, but the way things were going at the brewery, he'd likely be leaving before he could afford to. The money had never been ours, of course, and none of us had ever had expectations of being "rich," but knowing the Stroh trusts were there had given us the space to dream of the lives we wanted for ourselves.

"Hold on a second, Bill," I jumped in. "Elisa is not the mother of our father's children. That's got to count for *something*."

"No, she's not," Bill said. "But, you see, when these trusts were written, back in the forties, divorce wasn't common. Your grandfather assumed that 'lawful wife' meant first and only wife. Times and circumstances have obviously changed since then. The documents, however, make no distinction between first and second wives. Or even third, for that matter."

Whitney shifted restlessly in his chair, his upper lip beading with sweat. "Bill, are you saying that Elisa will likely get the biggest piece of the pie?"

Bill looked at all of us with sympathy. "At my urging, your father and Elisa are currently negotiating a *post*nuptial agreement. But . . . well, your father has no leverage, really, at this point. Elisa and her attorney have already rejected several generous proposals. So . . . you three should be advised that Elisa, now that she's legally married to your father, can pretty much write her own ticket."

"And you can be sure she will," muttered Bobby.

"Afraid so," Bill said.

I put my pen down on the table and pulled on my coat. "Are we finished?" I asked.

Everyone nodded.

"Merry fucking Christmas," said Whitney to no one in particular.

CHRISTMAS, 1974

(by Eric Stroh)

My father's house was brimming with Christmas spirit. We'd come straight from the meeting at the bank, stiff from cold. Handel's *Messiah* was piped in from hidden speakers in the walls, and a handsome trimmed tree brightened the living room, the floor beneath blanketed with festively wrapped presents.

"Bah, humbug!" said my father with a big grin as we sat down. "How about some eggnog?"

"Sure," I said. I needed a drink. At least here I could count on more whiskey than nog.

"Elisa will be back any minute," said my father cheerfully as he left the room.

"She's dropping off some presents at her father's house."

Bobby, Whitney, and I gave each other meaningful looks. This was the first time we'd all be together as a "family," and we'd agreed in the car not to let on about the meeting. Bill had told us that my father knew we'd come down and would be calling for the report, and I guessed this was the reason for my father's uncharacteristic cheerfulness. The meeting, after all, had gone off without a hitch.

Whitney suddenly stood up and stamped the snow off his loafers onto the Berber carpet, as if marking his territory before Elisa returned.

Bobby laughed at his younger brother. "Nice. Now there's going to be a big puddle in the middle of the goddamn carpet. Dad's going to *love* that."

"Better there than on my shoes," Whitney said. But a moment later he picked up the chunks of snow and carried them into the kitchen. My father's benevolent mood, he knew, especially toward him, was as changeable as the wind.

Beautiful objects adorned every surface in the room: an antique partners desk stood in a bay window with a gilt-framed painting by Gari Melchers on the adjacent wall. Tasteful patterned fabrics covered the upholstered furniture. Eighteenth-century walnut side tables held needlepoint coasters for drinks. The tree sparkled with old family ornaments and colored lights, just as it had when we were children.

Back in those days, before the advent of video, my father would draw the curtains, on Christmas Eve Day, to project 35 mm films of all the Christmas classics—*A Christmas Carol*, *It's a Wonderful Life*, *White Christmas*. The living room would be packed with spectators: my parents' friends, their kids, and, of course, the four of us. The adults drank Bloody Marys all afternoon, forming a chatty line at the tabletop bar while my father switched reels midmovie.

The next morning, Bing Crosby's "White Christmas" would

be playing on the stereo when we came down to investigate our stockings. Charlie and Bobby would get toothpaste, toothbrushes, deodorant, pens, and a jar of macadamia nuts, contents that left me baffled by Santa's odd sense of practicality. Whitney and I got dime store toys with the price tags still attached, tubes of toothpaste, and packets of pencils. Most years my father slipped into my stocking a Cuban cigar, which I'd smoke with him in the late afternoon, after the guests had gone home.

My father came in with the eggnog and placed the glass next to me on one of the coasters. "Ho, ho, ho, Merry Christmas!" He gave me a goofy smile. "How'd you like the tree?"

"Very nice," said Bobby. "When did you put it up?"

"Last week, before all the snow. I haven't had a real tree in years."

For about a decade, after the divorce, my father had pretty much given up on everything, even Christmas, making due with a small tabletop tree. He even gave away his entire 35 mm film collection—hundreds of rare prints. But in spite of everything, he'd managed to stay away from the bottle. Now, though, along with the tree, he had brought the drinking back into his life.

I spotted a small aquamarine Tiffany box under the tree. "Who's *that* little blue box for?" I asked pointedly, remembering Christmases past when the Tiffany box had been for me.

"That's for my sweetie," my father beamed.

It took me a moment to register that he meant Elisa. I glanced around the room and noticed that the family photos were now mixed in with shots of her. I looked over at Bobby, who was trying to suppress a laugh.

Whitney came into the room downing a Pepsi. "I like that painting of the pheasants in the kitchen," he said to my father. "Where'd you get it?" He sat down on the sofa, nervously crunching the half-empty Pepsi can.

"Out in Jackson," said my father, putting another log on the fire. "At a wildlife art gallery."

Whitney put the Pepsi can down on a coaster. "Outstanding. Maybe we can go there next week, see what else they've got . . ."

Rising from the fireplace, my father gave Whitney a disingenuous smile. "Trouble with *you*, Whit, is you've got champagne taste, on a beer budget." This was his favorite line.

Whitney looked as if he'd just been slapped.

Coming on the heels of our visit to Bill Penner, my father's gloating seemed almost sinister. The three of us could put on a good face for only so long.

Just then the wreathed front door opened with a rattle, registering somewhere between festive and frantic.

"Hello?" Elisa's voice called from the front hall as a gust of freezing wind brought the temperature in the house down several degrees all at once.

"Close the damn door, Elisa!" shouted my father. "Then come on in to the living room. We're in *here*."

The door slammed shut, and a moment later Elisa came in, ruddy-faced and wild-eyed, her glance shifting self-consciously from face to face.

"Look, everyone's here!" she burst out, giving me a big-breasted hug before turning to shake hands with Bobby and Whitney, who stood up to receive her. She wore lumberjack boots and an oversize down jacket that she absently tossed onto the rug. "How're things on the other side of the pond?" she said, turning back to me.

"Not bad." I smiled, pretending, as she was, to be "comradely." It was going to be a long week out in Jackson, I realized.

"Hey, have the police found out who *burgled* you?" Before I could even answer, Elisa roared with laughter at her own quip while my father gazed at her admiringly.

The burglary had happened while she and my father were in London, making it an easy conversation topic over several dinners. She'd asked me this same question, and laughed in the same way, a couple of times before.

"Not yet." I sipped my eggnog. "But, you know, they're searching high and low."

We turned the corner onto Lakeland Road, the snowplow just ahead of us spewing salt and snow in all directions. Every door on the street was wreathed. Some had Christmas lights strung across the shrubs, or fixed around doorway arches, that were coming on in the fading light. When we were younger, back on Provencal Road, my mother would drive us to other parts of Grosse Pointe to see the showier displays of Christmas lights and lawn decorations, usually at the "new money" houses along Lakeshore Drive.

"I hate this house," said Whitney bitterly as we pulled into my mother's driveway.

His palpable anger needed somewhere to land. I made no effort to correct or appease him.

"You could always stay at Dad's house," Bobby answered, deadpan, as he parked the car. "With your lovely stepmother."

For a few moments we just sat there in the rental car, a bright red Ford Probe, the windshield wipers scraping loudly over the ice that had frozen to the glass. The house looked gloomy, all the rooms dark except for the library, where I knew my mother sat reading. She had no Christmas lights outside, and the lights on the tree in the living room hadn't yet been turned on. She'd never gone in for frills. When we were kids, on an excursion to buy Whitney a fishing rod, my mother had told him adamantly, "Nothing fancy, just a stick and a string."

No one made a move to get out of the car. "We haven't talked about everything that just happened," I said tentatively.

Bobby turned off the car, and the wipers stopped mid-scrape. He jingled the keys. "Yeah, didn't want to be the one to start that conversation."

"I'll say it," Whitney interrupted him. "What a fucking turdfest! I mean, did you even see Elisa bring Dad that cocktail as we were leaving? She's up to a lot more than getting the money—she's trying to fucking kill the poor bastard!"

I shared his alarm, although I knew the drink wasn't Elisa's fault; nor was hastening my father's demise the same as causing it, though it felt good for the moment to have a scapegoat. From where we sat, it seemed we'd lost both our father and his legacy, and Elisa was an easy target. I'd often wondered if having money was more of a curse than an asset; and at that moment, money and death seemed hopelessly intertwined "Guess Dad and Charlie are in a race to die first," I said. "And Elisa's certainly getting all *her* ducks in a row."

"Your father married bar scum, Frances," said Whitney, as if announcing it officially. "Get used to it. Our family is like a *Vanity Fair* story on steroids."

Bobby and I laughed at the absurdity of this. But I felt gutted at the thought of my father drinking regularly again with Elisa. And after everything our family had been through, his getting drunk and eloping without a prenup seemed supremely selfish.

"It's just . . . very sad, that's all," said Bobby. He leaned back into the driver's seat with a sigh of defeat. "You're lucky you both live so far away. Watching this up close? It's going to be torture."

I exhaled two steady streams of smoke through my nostrils. "Believe me, I'd get on a plane back to London in a heartbeat if I could." Things had taken a turn for the worse in London, but . . . anything was better than this.

As I watched my brothers, I wondered how they would cope. Bobby gazed out the car window at the freshly falling snow, his expression one of resignation. Whitney sat smoking in the backseat, his face stamped with bitterness. Wondering where I would go, what I would do, I suddenly felt the deepest fatigue, the kind no amount of sleep might relieve.

Night had fallen. Still, we did not get out of the parked car. Soft clumps of snow floated down into the beams of the street lamps behind us, and every house on the street, with the exception of ours, shimmered with festive lights. I could hear the faint sound of singing and turned to see a neighbor's front door open and a group of carolers standing inside their warmly lit atrium. *Hark! The herald angels sing / glory to the newborn king!* They sang joyously to the crowd gathered around them, in the cold, cold night.

The sugary smell of Campbell's tomato soup filled the house as we went in at last. Bobby and Whitney went into the kitchen, and I retreated upstairs, feeling a headache coming on. My room still felt unfamiliar despite all the years of visiting my mother at this house, my old bookshelves, dresser, and desk arranged gracelessly against the polka-dotted walls, probably by the movers themselves when she first arrived here, never to be revisited.

I could hear Bobby's and Whitney's voices down in the kitchen, filling my mother and grandmother in on the meeting with Bill. I couldn't hear the words, but compared with the rise and fall of my brothers' distress my mother's voice was a low murmur.

I went into the bathroom, took two Advil, and looked in the mirror: there were swollen sacks beneath my eyes; my skin was pale with winter, translucent almost. My grandmother was right: my hair looked like hell—bleached and cut too short, a supercropped bob that succeeded only in rendering me utterly androgynous. I had even taken to slicking it back in the style of a man. The cut and color had looked chic in my mirror in England, but here, in my mirror back home, I could see only that I'd thrown away my looks, and two years of my life, to chase down a dream that now seemed as elusive as Santa Claus. An American woman, the next British art star? I laughed bitterly. Wanting to scream for all the wasted time, the squandered energy, the crazy hubris that had kept that dream alive, I grabbed my cheap hair and pulled until the tears came.

Homecoming

CRETE, 1997

(by Frances Stroh)

Detroit, 2000

"I know my brothers will like you," I assured my boyfriend. "They'll like that you're Russian, first of all, and they'll definitely appreciate your cooking."

We sped past Detroit, headed to Grosse Pointe on the Chrysler Freeway. We'd come from San Francisco, where we were living together. It was nearly our three-year anniversary, in fact. My family was gathering for an annual business meeting and, though Arkady had already met my parents on several occasions, I was taking the opportunity to introduce him to my brothers. Even Charlie would be there.

"I'll cook them a great dinner," said Arkady in his thick Russian accent.

Detroit had always looked bombed out, but now it seemed positively abandoned, with whole blocks of what appeared to be unoccupied houses. I knew many of the city's inhabitants had fled as late-stage industrialization set in, but I'd had no idea just how bad things had become. This was the legacy of

Coleman Young's twenty years in office. Young was a mayor who'd gloated as white businesses moved out, no matter how many black jobs were lost in the process. The desolation was the aftermath of shortsighted union leaders who had tailored their policies to bolster the faltering automotive industry and line their own pockets. And the consequence of the Big Three automakers themselves, who for too long had refused to build smaller, cheaper, more fuel-efficient cars to compete with their foreign counterparts.

Now, with the retreat of the automotive industry to the suburbs and abroad, all signs of manufacturing had finally vanished in Detroit, its smokestacks emitting no smoke. In many parts of the city, only footprints of manufacturing plants remained where the buildings themselves had been leveled. Defunct cement silos dotted the riverfront because the city was too broke to take them down. Crack cocaine, the cornerstone of the new economy, was a difficult commodity to tax.

Yet all the devastation had given birth to Eminem and the Detroit underground hip-hop scene, the first sign of life in a city that had been dying for decades. It had taken forty years, but Detroit now had its signature music scene again, one that reflected the hopelessness of the times, just as surely as sixties Motown had once mirrored the burgeoning black middle class's sense of abundance and upward mobility.

The arrow-straight freeway suddenly looped around the city at exactly the spot where the Stroh Brewery had once stood, its pinkish-red letters lighting up the sky. When the Chrysler Freeway was built in the sixties, the city had mandated that every building in the interstate's path be razed, except the historic brewery. If only Coleman Young had shown us the same respect and level of consideration; seventeen years and $300 million later, the stampede of tenants he'd promised at Stroh River Place had never materialized. Our office building was half empty, our apartments and hotel lost in foreclosure. My family's arc was eerily parallel to Detroit's; we'd boomed together and now . . . we were failing together.

"That's where Stroh's Beer was made," I told Arkady, pointing out the window.

He frowned at the weed-filled expanse of derelict land. "Unbelievable," he said, in a single word deftly acknowledging all my family's losses since the demolition of the brewery fifteen years before. "Too bad we can't drink Stroh's Beer with the dinner I'm going to cook for you," he added cheerfully.

Arkady was a world-class amateur cook, one of the reasons I had fallen for him. We met on Crete in the summer of 1997, shortly after I'd left London. He was sleeping on the balcony of a partially built hotel, under the stars. I was intrigued.

"I cannot sleep in rooms," he told me. "I need air—not *dust.*"

We cooked lamb on the beach and drank wine straight from the bottle. He did weight lifting with actual rocks, his Olympic muscles as chiseled as those of the ancient Greek statues I'd seen in Athens. When he brought me Mediterranean salt water to heal a sinus infection, I nicknamed him "The Bushman." He'd lifted the jar of murky water to my nostrils. "Inhale," he commanded.

For months, Arkady listened sympathetically to my family

saga. I told him about the family ski trip in Jackson just after Christmas, and how, with all the stress, I'd come down with the flu and had to extend my stay by three days, holed up in my hotel room and ordering French onion soup from room service. I convinced the airline to change my ticket back to London with no extra charges. My father appeared at the door of my room only once, holding his breath while dropping two Advil into the palm of my hand.

When Elisa, my father, and I checked out, my father told me, not a little sheepishly, that I would be responsible for covering the cost of my extra days—totaling about $900. "I agreed to pay for six nights," he reminded me.

Elisa stood at my father's side, looking on with what seemed a certain satisfaction.

Stunned, I pulled out my credit card. I had no idea how I would cover my rent back in London. I would move out of my flat, I decided, after using up my last month's rent deposit.

"It's a sad story," Arkady would say with conviction as we sat in this or that taverna, sipping Greek coffee. "This would never happen in Russia. *Never.*"

"But that's how I ended up in Greece, see?" I told him. "Because it's dirt cheap."

"A man's daughter should always come first. *Always.*"

Arkady's values and opinions were as solid as the rocks he lifted on the beach. When it seemed my family situation had toppled my flimsy house of cards, Arkady provided strength and sanity. Our days on Crete were filled with mountain hikes, dips in the Mediterranean, yoga, and the preparation of three organic meals. Arkady knew nothing about contemporary art and had little money; still, I felt nurtured when I was with him. As the weeks passed, the complexities of the art world and my family supplanted by a program of rigorous self-care, I developed a bodily sense of well-being I had never before known. My newfound sense of calm felt totally unfamiliar, and I slept through the night for the first time in years, often after listening to Arkady play Russian gypsy songs on his balcony.

Some days I felt perfectly at peace with the simplicity of my life on Crete, and yet there were other days when my body and my mind seemed at odds, when I longed for the intellectual stimulation I'd left behind in London, the city's frenetic sense of potential, even as the possibility of returning to that other life receded.

The change in me happened slowly, almost imperceptibly, like the leftover summer snow thawing on Crete's distant peaks. It happened incrementally, taking hold as we traversed the arid landscape past tiny white chapels dotting the hillsides; or strolled along beaches fiercely drifted by the sirocco winds, our sun hats skipping across the sand in front of us, our skin whipped and welted. Somewhere inside the casual rhythm of our days, I began to understand: not just that my entire life would change, but that it *had* to. I needed more substance, less abstraction; more space in which to move around emotionally and physically; and fewer hang-ups about money, fewer preconceptions. I would somehow find the way to depend on myself, rather than others. While in my work, I would benefit perhaps from a medium in which to deal with real events and real feelings—filmmaking, say, or writing.

After Greece, Arkady and I lived around the world for almost two years—New York, Kauai, Turkey, Kauai again—before settling back in San Francisco, where I had just finished my first short story. I had also been trading technology stocks and doing remarkably well, capitalizing on the crazy possibility of the dot-com era, while helping Arkady establish his yoga business. We were talking about getting married and having a baby, so . . . the time felt right now to bring him home.

The contrast with Detroit as we drove into Grosse Pointe was more striking than ever. Immaculate houses suddenly lined the wide boulevards. The upscale shopping villages were flanked with Noah's Bagels, Brooks Brothers, and Talbots. Mercedes and BMWs lined the streets, a clear sign that even the locals no longer bothered to buy American.

"Where's the Whole Foods?" asked Arkady after we'd passed the A&P market.

"There isn't one," I told him. "We'll have to drive out to Troy tomorrow for the turkey." I didn't attempt to explain the differences between old money and new money in suburban Detroit, or the fact that the old money Grosse Pointers would rather simply dine at their clubs than bother stocking their kitchens with locally sourced produce.

But Arkady planned to make a totally organic Thanksgiving-style dinner for my family the following day. It was March. By this point, my family rarely gathered over the holidays, and always without Charlie. We had to seize the moment, Arkady felt.

"I can't believe Charlie's going to be there," I mused. "He hasn't come home in seven years." Not since the Christmas just after I'd filmed the family piece.

"It will be great for you to see him," Arkady said. "Family is *so* important."

Like Detroit itself, Charlie had declined even further in recent years. After Bobby moved away from Dallas—first to Grosse Pointe, and then Tortola—Charlie got into crack and other hard drugs. Unable to hold down a job, he no longer even tried to. To keep him off the streets, my parents had appointed Bill Penner as Charlie's trustee. It was Bill who paid the bills: rent, utilities, food, travel; Bill who controlled where Charlie went and what he did; and Bill who had succeeded, until now, at convincing Charlie to stay away from Grosse Pointe. Legally, my parents and Bill could not keep Charlie out of a shareholder meeting.

Arkady ran his hand over his shaved head. "Of all your brothers, Charlie will be my favorite," he said. "I know it."

My mother sat at the kitchen table knitting while Arkady stuffed the goose (there were no organic turkeys to be had in Michigan in March) and I peeled potatoes at the sink. My mother's hair by now had gone completely gray, and her back was hunched from all the years of bending over backgammon tables and knitting projects—and she was bending over her knitting right now, in fact. The knitting needles clicked rhythmically to the hum of the dishwasher just as the dice and backgammon pieces had once clicked across sun-warmed corkboard at the club. Did my mother use the basic rhythm in these activities as an antidote to anxiety? I wondered.

"Charlie can be at the house for dinner," she said. "But he'll have to stay at a hotel. His seedy friends have put him up to this, I know. The poor thing, getting sent here by those grubs to get his hands on some money."

Arkady shot me an alarmed look, whereas I'd grown so used to my parents' callousness toward Charlie, I barely shrugged.

"Maybe Charlie, you know, just wants to come home," I offered. "It's been years since he's seen us." Charlie and I often talked by phone. Recently, he had confided in me that he had hepatitis C. My parents knew about this, too, but it had not changed their attitude toward him.

My mother shook her head. "No, dear. It's the meeting he's coming for."

Arkady seasoned the bird, his back to us. I loved his Old World values—good food, family bonds, and friends who mattered. He was deeply hospitable; he and his three brothers—talented musicians all—turned every gathering into a hearty celebration. How different they were from my family.

I hoped Arkady would understand the situation with Charlie—that my mother simply couldn't handle this son of hers; if Charlie stayed at the house, the cascade of guilty feelings would be too much. And so, it was her only defense, keeping Charlie at bay. Seeing how disturbed Arkady clearly was, however, I began to wonder if I should have brought him home for this particular weekend.

Among the only times we convened as a family, the annual Stroh business meetings were never happy occasions. Bobby, Whitney, and I always came home for the meeting, springing for expensive airline tickets only to be told just how badly the business was faring. After our acquisition of the bankrupt G. Heileman Brewing Company in 1996—landing another fifteen or so declining brands in our portfolio—sales had continued to plummet. In 1999, we finally sold our entire brewing business to Miller and Pabst, who divided up our labels like so many spare parts. The family was crestfallen, our 150-year brewing tradition gone, just like that. Miller Brewing bought our Henry Weinhard's and Mickey's brands. Pabst bought our forty or so remaining brands, including all the Stroh's brands, which they soon buried, even as its own label picked up a hipster cred it parlayed into record profits. Internet chat rooms had filled up with conversations among our many loyal consumers about what had happened to Stroh's Beer. No one knew.

Then, after the substantial business loans for the Heileman purchase had been paid off, the proceeds from the sale—just over a hundred million—went directly into the equities markets, only to take an instant nosedive with the "dot-bomb." And since my father had sworn Bill Penner to secrecy about the terms of his postnuptial agreement with Elisa, my brothers and I weren't even sure if a sliver of interest still remained.

Yet we still sprang for those airline tickets, if only because we knew these meetings would be the only chance we'd have to see each other.

The evening went badly. The goose Arkady had procured was not a success, its tasteless meat yielding no drippings for gravy.

"In Russia this would never happen," he said over and over as he sweated over the steaming bird carcass. "Farm-raised *garbage*."

Trying to help him save the dinner, Bobby's blond Louisianian wife, Cheryl, spent hours in the kitchen with Arkady. But it was no use; the American goose was just a total failure.

Charlie came into the kitchen as soon as he arrived. It broke my heart to see how irretrievably his looks had been disfigured by the drugs and hard living. "How's my princess?" he said, wrapping me in a big hug, as if entirely unaware of the shock I was registering. A moment later he was cheerful, chatting away with Arkady at the stove, bringing a can of Old Milwaukee Near Beer to his lips now and again. My mother had asked him to not have any alcohol, and he seemed agreeable enough about this. None of us would be drinking at dinner—"It would be cruel to drink in front of Charlie," my mother had said earlier that day—which was why my father and Elisa had opted out of the gathering.

When dinner was served, my mother sat at the head of the table in the dining room, her knitting pushed aside in a forlorn pile on the kitchen counter. She absently redistributed the food on her plate while Bobby, Whitney, and I discussed the agenda for the meeting the following day.

When the conversation lagged, we sat around the table staring at one another, unable to eat the tough, bone-dry goose meat. Whitney shifted restlessly in his chair, scuffing the wood floor. Bobby absently pulled on his mustache and looked at his watch. Charlie crossed and recrossed his legs impatiently, looking as if he might climb the walls. We all could have used a drink just then, but Charlie, he clearly *needed* one.

Whitney jumped up from the table first—to meet his friends at a bar. "If you'll please excuse me," he said in an official tone as he exited. "I'm expected elsewhere."

Bobby, too, stood up, and then Cheryl. "Thank you, Arkady," she offered sweetly in her Southern lilt. From the window, I watched them get into their car with somber faces and drive away.

Charlie dropped his napkin on the table and quietly went upstairs. My mother insisted I follow him. "Go up and make sure he doesn't take anything," she whispered. All evening, she'd been trying to head off some imaginary disaster.

Reluctantly, I stood up and, to appease her, I followed Charlie upstairs. I remembered a Christmas long before when he'd been accused of stealing forty dollars from my grandmother's purse. The family had made too big a deal out of it, talking in hushed voices about the pilfered money. He'd stolen from everyone else in the family as well, I knew, with the exception of perhaps me. From the very beginning, he'd always had my back.

As soon as I reached the top of the stairwell, I veered into my room.

Charlie had gone into Whitney's room, where I knew some of his old clothes still hung in the closet, along with some World War II Russian uniforms Whitney had collected.

"Franny, do you think Whitney'll mind if I take one of these shirts?" Charlie, who must have heard me, called out.

I walked through the shared bathroom into Whitney's room, where Charlie held up a blue oxford shirt on a hanger.

"Why not? Those shirts have been there for years. Whitney has all his clothes down in Florida."

"Okay, just checking." He sniffed the shirt. "Boy, you'd think Mom might get 'em cleaned once in a while, though."

I took the shirt and looked it over. "Maybe just keep it out overnight? The smell will go away."

"So you're staying here, huh?" he said without the slightest trace of resentment. "I'll stay down at the River Place Inn."

I pictured the shirt hanging off the bedpost in his hotel room, at Stroh River Place, the hotel that we'd developed, owned, and then lost in foreclosure, while Charlie slumbered underneath the sterile covers. "Yes," I said. "But I never sleep well here." I wished I could give him my room.

"Too bad. You could have joined me at the bar."

"I know." I hugged him. "Hey, I'll see you in the morning, though, at the meeting."

After Charlie left, Arkady and I cleared the table while my mother scrubbed the pots with a Brillo pad, steam rising around her exhausted face. The goose grease sat in a tomato-soup can by the sink.

"Make sure you never give Charlie anything for Christmas that he could sell for drugs," she warned. "Remember the time Bill Penner sent him the big-screen TV? He'd wanted one so badly, and then he sold it within a week." She shook her head sadly and went over to take a pie out of the oven. The party had broken up before dessert was even served.

I sat down at the kitchen table. "Obviously I'm going to send him a Christmas present," I said.

"Socks or underwear, though. That's what I do every year." My mother gave me this same advice every so often. I had yet to follow it.

"Come on, Mom, realistically how many pairs of socks and underwear does he need?"

"At least he'll know you were thinking of him." She dropped the pie from her oven glove onto the counter, knocking off a bit of crust. "Poor Charlie. He was never as smart as the rest of you. I spent years helping him with his speech, helping him learn all the things that came naturally to you and Bobby, and Whitney. He got a great deal of attention. More than all the rest of you put together, in fact."

"Yeah, well, Dad was pretty hard on him," I said. I flashed to Charlie quickly gathering up all the Indian beads on the bed as my father's footsteps came down the long hallway, toward his room.

"Genes play a much larger role in how we develop than environment. The more I look around, the more I see this." My mother sliced the pie and brought two plates over to the kitchen table. "When Gari and John married Susie and Lou, there just wasn't enough variety of genes."

For as long as I could remember, I'd been hearing my mother say the Stroh family's paltry gene pool came from those two brothers marrying two sisters—Susie and Lou, who drank martinis as if they were water. I myself leaned more toward the nurture side of the nature versus nurture debate.

"Dad picked on the defenseless ones, if you ask me," I said. "Bobby and I turned out okay because he liked us better, don't you think? I mean, we could stand up to him, for one thing." Bobby had once told me a story about throwing a baked potato at my father in the course of an argument, which caused my father to back down with a sort of respect.

My mother chewed her pie. "I still think nearly all of it's genes."

"Genes are very important," Arkady agreed. "They decide almost everything."

I looked at my boyfriend as he loaded the dishwasher and for a moment wondered what planet he was from. Of course that's what had attracted me to begin with—his utter certainty about things.

I took a bite of pie and thought about all the years of misplaced blame in our family. Whether it was the business going down the tubes or Charlie's demise, no one wanted to take responsibility; even our genes seemed little more than convenient scapegoats.

"You're probably right," I told my mother. I didn't have the heart to challenge her; the truth would be too much to bear.

"I know I am," she said.

DETROIT, CIRCA 1972

(by Eric Stroh)

At eight thirty the next morning, the family assembled in a meeting room in our office building at Stroh River Place to await the presentations. Forty of us—cousins, aunts, uncles, brothers, sisters, and parents, whether related by blood or by marriage—sat united in our morbid curiosity about the fate of the failing business, as well as our current or future financial interests.

The room hushed when Charlie entered, and you couldn't blame anyone. In the glaring light of day, he looked like a homeless man as he ambled through the room, his skin leathery and blotched, eyes gazing vacantly ahead, making eye contact with no one—a homeless man dressed up for church by community do-gooders. His large nose appeared to have taken over his face, a face cut into an apple and left to dry. The blue oxford shirt from Whitney's closet seemed to constrain his gestures, while the gray flannels and loafers he'd also taken from the house no longer fit his diminishing frame.

He pulled out the chair next to mine with difficulty, catching the white tablecloth and disturbing the glasses of water at each place. The acidic fumes of stale alcohol rose off his skin and breath.

I felt at once shame and compassion. This was my brother, even if many of the people in this room could hardly recognize him. Yet I was ashamed of what he'd become, as if his state somehow reflected on me. And recognizing my own wish to escape implication made me feel only worse. Waiting for the meeting to begin, Charlie crossed his legs and fixed his eyes on the overhead projector.

I wasn't sure how much he would be able to take in. Surely, he'd been absent for so long that the perils of the Detroit real estate market, the pension fund, and our biotechnology interest would mean little to him.

My cousin John, smart in a blue blazer and perfectly tailored gray flannel trousers, walked to the front of the room. Having replaced Peter as the board's interim head in 1997, he still acted as our CEO and chairman. He welcomed the group and, smiling, began his opening presentation.

"Allow me to state the obvious: the fact that we continue to meet in this building—in a city that is literally falling apart all around us—is not exactly good news."

Everyone laughed grimly. At dinner the night before, Bobby had characterized the situation with brilliant shorthand, "Replacing Peter with John was like rearranging the deck chairs on the *Titanic*."

Bobby had nailed it, though his metaphor left out any hint of responsibility. The fact was the Stroh Companies, Inc., board, as well as the family members who'd elected them, were responsible for this shipwreck. They had placed that fatal iceberg right in their own path.

"If we could pick this building up and put it down somewhere else, *anywhere* else," John continued, "we'd be okay. But for better or worse, our real estate holdings are in Detroit."

"We can thank the United Auto Workers for bringing Michigan to its knees," one of my cousins piped in.

"The unions certainly haven't helped us," John agreed. "If U.S. manufacturing had remained affordable for the Big Three, this wouldn't be the ghost town that it is."

Passing the buck, it seemed, had reached epidemic proportions in this family, and I wanted some accountability. "John, did we ever consider moving our headquarters elsewhere?" I asked. "After the Detroit brewery was closed in the mideighties, for example, and we had breweries all over the country?"

"No one had a crystal ball, Franny," John responded smoothly. "And uprooting management was not exactly an attractive option."

He went on to deliver the complexities of the bad news with a confident sort of complacency that seemed to belie the dire circumstances in which we found ourselves. The business—now a holding company called Stroh Companies, Inc. (SCI), with a basket of declining assets—would cease to exist within a certain number of years, and no one knew the exact number. Dividends would cease even sooner.

As the automotive industry had declined and businesses had moved out of the city, so our real estate holdings had plummeted. Stroh River Place had been our most disastrous investment, representing a loss of close to $300 million, some of which had been bank loans, triggering foreclosures on the hotel and the apartment building. As for the office building we sat in, at 50 percent occupancy it was now worth but a tiny fraction of what the family had invested in it.

John moved on to the next line item: Spain. The Spanish government was suing us for a tax they believed we owed on our Spanish brewing interest—Cruz Campo—which we had sold to Guinness in the early nineties. For years, we'd been fighting the tax in the Spanish courts; if enforced, it would have totally wiped us out. In the meantime, the SCI board had voted to pay a substantial portion of the company's remaining equity out to the shareholders.

"So save your money," John advised ominously. "We'll be dissolving the company soon."

My relatives masked their panic by taking studious notes or simply staring ahead with blank expressions. Like my father, most of them had enjoyed quarterly dividend checks for as long as they could remember. Some had been saving; some hadn't. In our branch of the family, with my father as the only shareholder, my brothers and I had never received SCI dividends, and probably never would. I looked across the room and noticed that my father and Elisa were just coming back from a cigarette break, giggling like teenagers, as if utterly oblivious to the bad news.

"Elisa and I are having a ball spending your inheritance," my father loved to tell me. I knew he wasn't joking, but now even his own money supply was on the verge of tapping out. The company had been imploding while quarterly reports sat in an unopened pile on my father's kitchen counter, he and Elisa busy spending the dividends before the checks had even arrived.

Next, John moved on to our biotechnology interest, Apex Bioscience, the company Uncle Peter had founded in the early nineties with $100 million of SCI funds. Just a few years later, John would sell this division to a German venture capital firm for nothing but a minor equity stake in lieu of proceeds. On an annual basis, the Germans initiated additional funding rounds in which we had to participate in order to keep our stake intact.

"Should we invest more funds in biotechnology?" John asked the group. The venture capitalists were raising money for a phase-two drug trial. We weren't ever asked to vote, or even to give our opinions. "They're asking for a minimum investment of two million dollars. We haven't committed at this level in over a year." Clearly John was stumped, and the board always deferred to the family, meaning to John and Pierre, cousins who'd inherited board seats from their fathers. For the very first time, John was turning to the family for direction.

We wrote our votes on scraps of paper and put them into a hat. I voted in favor of the investment. What difference did it make? We were going down so fast we couldn't count our losses. And while it was a gamble, Apex Bioscience was the only investment in our portfolio that had any potential at all. As my father sometimes said, "We should be a Harvard textbook case study on how *not* to run a business."

"Family-owned and operated since 1775," our beer labels

had boasted, but that was exactly our problem: in our system, fathers promoted their sons, sons who too often had neither attended business school nor proven themselves in other corporate settings. Running a regional brewery was a far cry from running the beer giant we'd become in the eighties and nineties. And . . . we'd simply blown it.

When Alan Bond, the Australian financier, had made the Stroh Brewing Company a billion-dollar offer in the mid-eighties, our family-run board turned him down. Within just a few years we were taking less than half that number from Coors, only to watch them back out of the deal. Our fire sale to Miller and Pabst in the late nineties had been the ultimate humiliation. At the last moment, Pabst backed out of the Stroh Brewery's pension liability and still we'd agreed to the sale, shouldering that liability ourselves.

Like most family members, I'd watched our business decline year after year from the sidelines. I could see now that many of these strategic pitfalls might have been avoided, had seasoned professionals been at the helm. Family-owned and with a family-run board, we'd been in way over our heads once we became the third-largest North American brewer. Anheuser-Busch and Miller—number one and number two, respectively—were publicly traded companies with diversified boards and shareholders who kept a sharp eye on strategic decisions. Stroh shareholders, unless they happened to sit on the board, had no voice at all. These meetings were nothing more than elaborate smokescreens designed to foster a sense of agency, however false, among the rest of us.

When Uncle Peter retired in 1997 some of us had pushed for hiring outside talent instead of appointing yet another Stroh family member to run the business, but with the family divided on the question, the status quo had prevailed, and now, yet again, it seemed we were paying the price.

Suddenly, Charlie spoke up. "John, I have hepatitis C and, well, I need financial support for my medical care and living expenses."

The room gasped: the drug addicts' disease. Or one of them, anyway. Everyone looked at Charlie, stunned. "So how would I go about getting a guarantee of financial support?" he continued.

John, who had married a wealthy woman, cleared his throat. "Those of us who need support," he intoned, brows raised in seeming concern, "will have to turn to our parents."

My father shot Elisa a theatrical look of alarm. She grinned wanly and patted his knee. My mother did not lift her eyes from her knitting. Charlie crossed his arms and sank into his chair. I was proud of him, though, for speaking up. Many of us had watched our parents' generation spending the family wealth all our lives, but only Charlie had showed the courage to ask, "What about me?"

I leaned back into my seat and stared up at the ceiling. Every time I attended a Stroh business weekend, I swore I'd never attend another one. I felt forever caught between wishing we'd just hurry up and lose it all, and hoping we could save ourselves. But John's response to Charlie's question seemed to make it perfectly clear there would be no road back.

. . . .

At lunch, everyone steered clear of Charlie. I walked over and sat down in the seat next to him.

"Hey, Franny," he said, his mouth full.

"How's the pasta?" I asked, knowing the food would be perfectly tasteless.

Charlie pulled his napkin onto his lap when he saw me do it. "Not bad. Better than that goose last night." He laughed. "Hey, I like Arkady, though. You know? He's a good guy. I just couldn't stick around after dinner. Mom was too nervous. I can't stand being around her when she's like that."

I spotted my mother at a table across the room with Nicole and Pierre. "What did you think of the meeting?"

"Nothing new, right?" said Charlie. "This ship's been sinking since I can remember."

"No kidding."

"Anyway—sorry about the goose," he said. "Arkady worked like hell on that meal. It really wasn't his fault."

"He likes you, too, Charlie."

Later, back in San Francisco, Arkady would tell me that Charlie had pulled him aside and said, "You take care of my princess, okay?"

MAPLE LEAVES, 2000

(by Eric Stroh)

I left Michigan on a bone-chilling morning. My mother took us to the airport, and Arkady flew on to New York to see his brothers.

My mother waited with me at the gate for my flight to San Francisco. She sat knitting a baby blanket for her friend's grandchild and talked about how awful the weekend had been with Charlie there. After the meeting, she told me, he'd gotten drunk and made a scene in the lobby of the River Place Inn, shouting at the people checking into their rooms about Jesus and redemption.

"It's just dreadful, how he acts when he's been drinking," she said, her knitting needles clicking away in a steady rhythm. "It's terribly embarrassing. Poor thing, he just can't control himself. It's like Jekyll and Hyde."

I listened as she knitted herself back together with absolute truths, the baby blanket draping over her lap as it ever so slowly grew. She had reconstructed the weekend so seamlessly,

I almost found myself believing it, too: Charlie was the problem.

The day before, the hotel had called my mother about his minibar bill, which the family company had a policy of not covering.

She had called Charlie immediately. "Charlie, your minibar bill is a hundred and sixty *dollars*. How can you possibly drink that much in two days?"

"YOU'RE A FUCKING BITCH!" he'd shouted, my mother told me. The rest she wouldn't repeat. My mother had hung up the phone, dialed the front desk of the hotel, and given them Bill Penner's number.

"He was drunk," she explained now as the gate filled with people. "Look, he's very sweet when he's sober, but . . . when he's *drunk*, it's simply terrible."

Seeing how drained my mother looked, how the events of the weekend had worked her over emotionally, I pitied her nearly as much as I did Charlie. I only wished she could see the truth: that all the brushes with the law, the boozing and the drugs, the slow-motion suicide—these were Charlie's cries for help. Even now. He was challenging her. Challenging her to love him. In spite of everything, he was still her child.

The crowds milled past us on the way to their gates. Listening to the boarding announcement for the first-class passengers on my flight, I knew I had about ten more minutes before it was my turn to board, and I could put this weekend behind me.

My mother suddenly tensed and dropped her knitting. "There's Charlie," she said in a hollow voice.

There he was, not twenty feet from us, trailing through the airport, glancing around absently, unshaven and seeming almost lost. His ruined face struck me now as open and curious—an expression of almost childlike innocence.

Startled, we instantly looked away—perhaps because we were tired, or because he wasn't the person we'd spent the morning re-creating in our thoughts. Standing there as a physical fact, he was just a person in the world with a hangover and a plane ticket—a brother, a son. And yet someone to be kept at a distance. My mother and I looked at each other, ashamed.

"Let's say good-bye to him," I finally said, halfway out of my seat.

My mother's hand clenched my forearm. "No, Frances. I can't take anything more. *Please.* Just . . . let him go."

I turned my head, stricken, and watched Charlie wander off to his gate, disappearing into the crowd. He glanced our way but his eye never caught mine. Had he seen us? Perhaps it didn't matter. The moment was lost. I gathered my purse, boarded my flight, and just before takeoff vomited in the jet's toilet.

It was the last time we ever saw him.

Flowers

CHARLIE STROH, 1970

(by Eric Stroh)

San Francisco, 2003

The phone rang out into the sun-filled room and I answered quickly, taking care not to disturb the baby next to me on the sofa. It was Charlie's voice on the line.

"Franny! I'm at the Hyatt in Dallas," he said. "With my new lady friend. Here—talk to her."

I lay on my side with the phone balanced at my ear. My tiny, blue-veined son slept in the hollow of my armpit. Suddenly, an unfamiliar female voice greeted me, and we exchanged friendly words before Charlie returned to the line. "Her old man's threatening to kill me," he said with an unmistakable thrill in his voice. "He's circling the parking lot in a Hummer—with a shotgun in the front seat."

"Maybe you should call the police," I said, my body tensing. Mishka squirmed in the crease of my arm and began to cry. I wondered if Charlie was just high, or whether there was in fact a real threat. Arkady was cooking borscht in the kitchen, and I nearly called out to him.

"Did you get the flowers?" Charlie suddenly asked. "Those were Roxanne's idea."

I looked over at the flowers sitting on a table, a mélange of pink roses, baby's breath, and daffodils. "I love them," I told him. "You're the only one who sent me flowers."

"Yeah? Well, I can't wait to meet my nephew," said Charlie, his voice bursting with pride. "I'll come visit you in June. I promise."

Though Charlie had never come to see me in San Francisco, we talked on the phone reasonably often. A few years before I moved here, he told me, he'd taken a floor of rooms at the Fairmont Hotel and thrown a raging party. Good times.

"I'm going to hold you to that, Charlie," I said, before getting off the phone, almost believing he would come. "Okay? So I'll see you in June, then."

Charlie had taken to migrating from one Dallas hotel to the next, moving on when he'd worn out his welcome. Bill Penner, who by now had become a father figure for him, not only paying the bills but also offering him much-needed guidance, was the only one who knew his whereabouts on a consistent basis.

"Charlie's a damn nice guy," Bill would say to me over the phone when I'd call him to check on Charlie. "Shame he doesn't have a better relationship with your dad."

My father's attitude toward Charlie had only further soured over the years. In his mind, this second-born son seemed to represent all his own failures as a human being. "I wish I could push a button and just make him disappear," my father said more than once.

Both my parents had completely detached themselves from Charlie. "I *had* to create distance," my mother said. "It's the only way I can live." It was three years since she'd spoken with him—since the family meeting.

June came and went in a blur of sleepless nights, with Mishka feeding every two hours. In my delirium I would flash to his April birth, the midwife coaxing me out of our bathtub and onto the bed as I labored, coaching me in her heavy German accent to *push*. Arkady brought aromatic herbs and berries from the kitchen, distracting me from the agony. And the breaks between contractions had felt glorious, an intense endorphin high—what I imagined heroin must be like.

When Mishka arrived, Arkady and I looked at each other and laughed out loud, then cried. The midwife placed our baby's body on my stomach, and my love flared, fierce and unconditional. Then she sewed me up by the light of our living room lamp with the shade removed while Mishka blindly navigated my chest.

I never remembered to call Charlie about his visit; maybe I'd never really believed it would happen. We could go years, after all, without seeing each other. The distance had come to feel . . . normal.

And then, one windswept day in July, the phone rang. All day something had felt off—a weight, a fatigue, something beyond the usual. Mishka was napping in the bedroom, so I picked up in my writing office.

"Frances," said my mother, her voice tight. "How are you?" It wasn't a question.

"Fine," I said.

Then she told me, her voice artificially calm, matter-of-fact, that Charlie was dead.

He'd been staying at an Embassy Suites Hotel in Dallas, on the tenth floor.

According to the police report, Charlie had called the front desk of the hotel shouting for help. Someone was chasing him, he'd insisted, trying to kill him. The hotel sent the police up to his room to investigate. When two policemen arrived, they found a "paranoid man who believed he was being pursued." Opening the closets for Charlie, they reassured him that he was alone in the room and safe. Charlie seemed calmed. One of the policemen patted him on the shoulder, told him to "take care," and then they left.

The police took the elevator down to the lobby, spoke for some minutes to the hotel manager, and then exited the hotel. At the very moment they stepped outside, a large mass tangled in a white sheet fell from the sky, landing onto the pavement in front of them with a horrible thud. They looked up. Torn bedsheets, tied together and fashioned into a makeshift rope, dangled from the balcony of a room on the tenth floor.

My mother sat knitting in the living room of my apartment, her face sagging in a way it never had, the eyes uneven, the skin of the cheeks thickened. Mishka was draped across her chest, asleep. "It wasn't suicide," she said, her knitting needles clicking. "He was trying to escape."

"Who said it was suicide?" I asked.

We'd all assumed, on the basis of the police report, that Charlie had been attempting to escape his imaginary pursuer by making a rope out of bedsheets and lowering himself off the balcony.

"One of your cousins," said my mother. "But it *wasn't*."

I watched my mother cradle Mishka with her arm as she knit him a blanket. Her first grandchild. This visit, long planned, had taken on a dual purpose, and she looked at once totally wrecked and wholly content.

"I know it wasn't," I said.

I was stoking the logs in the fireplace, thinking of that line apocryphally attributed to Mark Twain, "The coldest winter I've ever spent was a summer in San Francisco."

Arkady had driven the three of us to Big Sur the day after receiving the news. In the evening I'd walked with our new baby down some stone steps to a hot spring, resting him inside his Moses basket, and looked up at the sweeping dome of stars. Waves crashed on the sides of the cliffs below, one after the other. It all felt vast—the sky, the rocks, the ocean—vast and oppressive.

"The service will be in Grosse Pointe in September," my mother was saying now. "He's being cremated."

I didn't want to know any details about the body, what condition it had been in, how it had been shipped back to Michigan. Nor did my mother ever think to mention these things. Bill Penner, I imagined, had handled everything.

"Helluva nice guy," Bill said when we spoke just after I received the news. I could tell he'd been crying. "I'll sure miss him."

"So will I." I myself had been unable to cry. I kept telling myself we'd always known something like this was coming. But the guilt I shouldered now was, in truth, too much to feel.

"Have you spoken with Whitney?" asked my mother.

I crossed the room and sat down in a chair. "He's holed up in Missoula," I said. "Getting a couple of summer school credits to finish college."

"He was so upset he had his girlfriend fly out to be with him," said my mother. "He's taking this very hard."

For all his bravado, Whitney could be deeply sensitive. I remembered how we'd both cried quietly in the car the year before, just after we'd visited Uncle Peter, who'd been diagnosed with a brain tumor.

The three of us had sat together in Peter and Nicole's living room while Nicole talked on the phone upstairs. Outside the bay window the golf course burst with savage green, cruel beyond measure in its vibrant health. Peter sat quietly in his tweed jacket and gray flannels, slightly shrunken, unable to make the usual conversation. The poor man was stunned, having received the terrible news only days before.

"You're going to be okay, Peter," I told him, putting my hand on his, echoing his words to me all those years before.

He looked at me with those warm, small eyes for the last time. "Thank you, Franny," he said, managing a smile.

Four months later, he died.

Mishka gurgled and my mother adjusted him gently. I poked at the logs to get the embers going again.

"And Bobby? Have you spoken with him?" my mother asked.

"I called him as soon as I heard," I said. I didn't tell her, though, what Bobby had said: "Charlie was a fool right up until the end." I knew his callousness toward Charlie was only a defense, like my father's; still, it hurt to hear it. Bobby's handling of the situation, like my parents', disturbed me almost as much as my own behavior did. We'd all distanced ourselves, rather brutally, it seemed to me. Charlie's neediness reminded us of our own frailties, and we'd hated him for it.

"Poor Charlie," was all that my father had said when I called him from Big Sur, a whole lifetime of regret in those three worn syllables.

My mother put down her knitting. Her gaze drifted to the crackling logs in the fireplace. She rested her hand on Mishka's back and closed her eyes.

I went over and spread a blanket across them both.

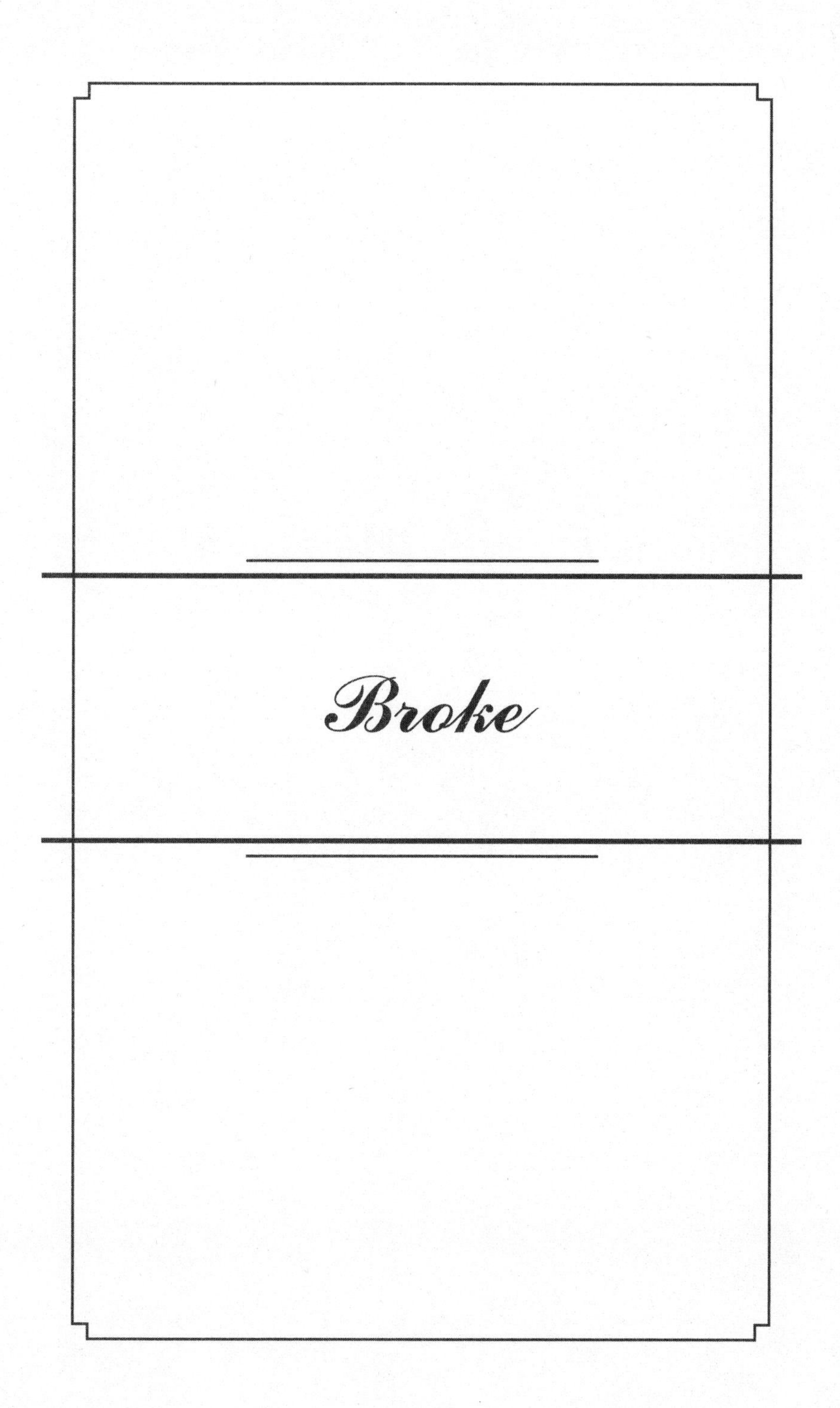

Broke

WYOMING, 2007

(by Eric Stroh)

Grosse Pointe, 2008

My father, five-year-old Mishka, and I sat at a table having dinner at The Hill, the restaurant where you ate when you weren't at the club. Dark paneling, starched linen tablecloths, weighty cutlery. A breadbasket heaping with gnarled rolls that resembled tightly wound fists. We'd flown into Detroit for twenty-four hours on our way from New York to San Francisco.

"I can't sleep," my father was saying, visibly irritated. "I'm so damned wound up." He was drinking club soda. Since his divorce from Elisa, his stomach couldn't take the alcohol anymore. Instead, he took painkillers.

He must have been reading his mail. Usually I was the one who filled him in on the grim news that arrived in the quarterly reports from the family holding company. The most recent had warned that dividends would be entirely eliminated in a few months, leaving my father to live on nothing more than his small pension. Sixteen years after our final listing in *Forbes*, the coffers were empty.

"Dad, you'll be okay," I reassured him. But we both knew he had nonnegotiable financial obligations both to my mother and to Elisa. To cover these, he likely would be forced to sell his house in a real estate market depressed beyond recognition by the bankrupt automotive industry, burning up what little he had left in those family trusts that Bill Penner had agonized over twelve years before. My mother would likely forgive the debt, but we were not counting on Elisa, who had left him for another man less than a year ago, to do the same.

My father slathered a roll with butter. "I can't remember the last time I heard good news," he said. "Whenever I open a goddamn letter it's always doom and gloom."

All the fears I'd ever felt about my own future paled in comparison to what I imagined he must be feeling about his. My father was not equipped, I knew, to live without a substantial income. I wondered if he would spend the rest of his life regretting his choices, or perhaps even wishing that he'd been born into another family, one that hadn't taken such good care of him, up to now.

That meeting in Bill Penner's office twelve years before had been the best thing that had ever happened to me. Since then, through my active investing, I'd parlayed a small investment account into a sizeable nest egg. I still had my income from the Detroit real estate trust, but I'd been informed that would soon end. No matter. Striving for something gives life its meaning, regardless of whether we succeed or fail. The problem was, my father had never *had* to strive for anything.

Looking over at Mishka, who was playing with the salt and pepper, I felt deeply grateful this would not be his fate.

My father sipped his club soda. "So—how's your book coming along?" he asked, suddenly upbeat. "Can I be a character in your great American novel?"

I was writing a novel set in the late-nineties New York art world, with an artist protagonist whose family had lost their wealth. I had stopped making installations years ago. Now I just wrote about artists. "Sure," I said, to please him. "Or maybe I'll start a *new* book—with you as the main character."

"Pipe-smoking old kraut?" he chuckled. "That kinda thing?"

I laughed. "Exactly."

Mishka was lining the silverware up across the table like a snake.

"He's a good-looking boy," my father said, watching Mishka with admiration. "Hope he turns out better than some of *my* kids did."

Charlie, he meant. Ever since stammering the words "poor Charlie" after we'd gotten the news, I hadn't heard him mention Charlie except by indirect reference.

"He will, Dad. Mishka will turn out fine."

"Only it's harder to control them when they get older," he said wistfully. "When you don't even know who their friends are."

My father ran his hand like a big spider up Mishka's arm, smiling like a kid himself. Mishka shrieked with laughter and came around the table to hug his grandfather, his blond hair the very color my father's had been before it turned silver.

"Women—that's the other problem," my father went on. "They're after only one thing. Pick you to the bone."

"Mom didn't," I reminded him.

He lit a cigarette. "Fair enough," he said.

I'd spent the final year of my father's marriage agonizing over whether or not to tell him about Elisa's affair. She'd asked my father for $6 million to build a boat and sail around the world with her new "friend," and my father had taken this at face value. But when Elisa attended Arkady's yoga workshop in Mexico, one night at the bar she'd bragged about her liaison. "Eric would do *anything* for me," she told the group between gulps from her Corona. "Because he doesn't know about sailor boy."

My own marriage had been failing at the time. Financial stress had been hard on the relationship, as had trying to run a business together. And taking care of a new baby, I hadn't had the bandwidth to grieve Charlie properly. I feared sometimes I might be carried away by the torrent of my own anger and sadness, and distance became my new mode of survival, all those unresolved feelings having calcified into a wall that kept everything out, even happiness. The withdrawal from the marriage happened over a period of years, and during that time both Arkady and I hoped the distance was only temporary. Perhaps it had been easier to focus my attention on my father's problems.

Then one day, when Mishka was almost four, I called my father from an airport and told him that Arkady and I would be separating. "We'll remain good friends, though," I added. "Just as you and Mom always managed to."

"Elisa's moving out, too," my father said. He soon filed for divorce, at Bill Penner's urging, and the round-the-world boat trip was canceled.

We left the restaurant, my father, Mishka, and I, and drove over to my father's house, the lush summer lawn out front in full bloom. For the first time in thirteen years, there was no need to check my father's driveway for Elisa's car, hoping she would be out.

My father had just finished remodeling. "How do you like it?" he asked when we came inside.

I peered into the living room from the hallway and spotted a real skeleton sitting in an upholstered armchair. My father had always been fascinated with skulls and bones, and often had a skull sitting on a bookshelf. "It's beautiful," I said, reminded suddenly of his solemn statement about Elisa eight years before: "If anything ever happened to her, I wouldn't be able to go on living."

A few months later, my father, by now suffering from diabetes, noticed an infection in his leg. An infected sore, hardly unusual in a diabetic, could have been treated easily enough. But he decided to let it go—for months. Did he suppose he had lived long enough—or that living without money was a fate worse than death? Perhaps systemic gangrene seemed a better choice than poverty.

Calling me late one night in San Francisco, just to talk, he seemed uncharacteristically at peace. We discussed the weather and my book, avoiding entirely the subject of the business. I told him how much Mishka loved the toy electric car he'd sent for Christmas. For once he didn't tell me to "speak English" when I said Mishka instead of Michael, our running joke. We laughed a lot. He seemed almost high, in fact. Unfettered. Something was wrong.

"Dad, it's late . . . what are you doing awake?"

"I love you, Franny," he said then.

I told him I loved him, too, and clung to the silence that followed, before he hung up, as if everything we'd ever wanted to say was there in that pause.

When he collapsed on his bathroom floor two days later, Ingrid, his housekeeper, took him to the hospital. He forbade her to call anyone in the family.

One week later, he died alone at four o'clock in the morning.

SELF-PORTRAIT OF ERIC STROH, 2004

In my father's house just before the funeral, I noticed that the skeleton that had been sitting in the armchair was gone. I went and sat in its place, taking in the scene for the last time. Soon everything in the room would change. Several antique handguns sat on top of the mantel. Rare books filled the bookshelves. The eighteenth-century celestial and terrestrial globes flanked the fireplace, and I was transported back to that Easter weekend long ago; I heard my father's shoes crunching the Manhattan pavement, felt his warm, protective grip on my hand, inhaled the exhaust from the taxis passing us on Park Avenue. My father looked down at me and smiled, just before we crossed the street. "Having fun, Minuscule?" he asked.

Clutching his hand more tightly, I told him I was.

. . . .

When my family filed into Christ Church—all stone inside with mahogany pews—I kept my eyes down. The damp air seemed to have an electric charge. Feeling hundreds of eyes trained on us as we took our seats in the front row, I clung to the piece of paper on which I'd written my reflections on my father's talent and character—a crystallization of all his best qualities, but an honest one, one that acknowledged his difficult side as well.

As the minister spoke, I thought about my father's ashes, which would go into a wall with a bronze plaque inside Christ Church's rose garden, the same wall where Charlie's ashes had been bricked in five and a half years before. When I visited Charlie's place in the wall, I left flowers. For my father, I would leave Cuban cigars.

And then it was my turn to speak. I walked over and took my place at the pulpit. The rustle of programs, with my father's photograph on the front, filled the church. I looked out at the crowd. Charlie and my father were not among them, and never would be again. With difficulty, I began to read the piece I'd written on the plane. I talked about my father's talent as an artist, and how photography had been one of the only ways he knew to connect with others. I talked of how difficult he could be at times, but also about the sensitive, generous man who'd been hiding beneath that gruff exterior. I talked of his loneliness, and how, like any of us, he'd only wanted to be happy. His heart had been like the shutter of his camera—opening wide for an instant, allowing the warmth of his spirit to escape into the room, however briefly.

My eulogy done, I looked up and spotted Elisa in the fourth row, her cheeks wet with tears. Had she recognized those same admirable qualities in my father? And had he felt this? Many of the people in the church had not reached out to my father in years, not since his marriage to Elisa. People were complicated. We failed ourselves, and each other. But we were all here now.

And so, as Elisa's eyes met mine, I smiled.

"Open a few windows," suggested Bobby. He, Whitney, and I had driven over to my father's house to sort through his collections. "Let's get some air in here." Bobby's dark mustache had flecks of gray, and he wore the rope anklet that marked him as an islander. At the funeral, his eulogy had made light of my father's punitive parenting style, recalling how, when we were kids, my father had jestingly referred to our house as "*Stalag* Stroh," after the term used for German prisoner-of-war camps during World War II.

Whitney opened the sliding glass doors leading to the terrace and went out to smoke a cigarette. He was in the midst of a drawn-out divorce and looked tired, his handsome face sagging in the dim winter light. He settled into the chair where just a few months before my father had sat smoking a cigar and balancing Mishka on his knee.

I wandered through my father's house. Eighteenth- and

mid-nineteenth-century English antiques mixed with sumptuously upholstered chairs and wildlife paintings in gilt frames. I had long ago left behind the world of monogrammed sweaters and award-winning gardens that my father's classic taste evoked, though my own rooms in San Francisco were peppered with English antiques I'd inherited from Stroh relatives over the years. Never feeling I had the money to properly decorate, I'd been glad to take the furniture, though in my fantasy I lived in a house full of Gerhard Richter paintings, Eames chairs, and sleek sectional sofas.

In my father's library, framed photographs of all of us at various life stages were mixed in with the books. I felt grateful that Charlie wasn't there to witness the absence of any shots of him as an adult anywhere in the house. Our studied smiles in the images covered over something else—a weight, an implicit knowledge that soon it would all come undone. And yet I felt a good deal lighter now. In all my weeping for my father over the previous days, I'd finally found the way to let Charlie go, too. Grief was perhaps undifferentiated.

I spotted the Dickens set—the one my father had purchased on our trip to New York when I was six. The gold-embossed spines of the volumes shimmered in the sunlight on his bookshelf. An apparition, a memory long locked away in a treasure chest, beheld once more. The house and most of its contents would soon be gone, just as the brewery was. We'd somehow allowed ourselves to be pinned into place by these things; and in our search for freedom, some of us had self-destructed.

I walked into the living room to find Whitney lining

up my father's six most valuable oil paintings, leaning them against the back of the sofa. He'd brought them up from the safe in the basement and had removed their protective, acid-free cloth covers, the same gilt-framed paintings he had considered removing from the house just after my father died for fear their value would create an enormous tax burden to the estate. Later, after receiving the appraisal report from DuMouchelles, we realized his worries had been pointless.

"You choose first," I said. I had my eye on a rain-drenched Paris street scene by the postwar impressionist Edouard Cortès, my heart suddenly racing as if I were bidding at an auction.

I was relieved when Whitney tagged the other Edouard Cortès—of a Paris flower market.

I chose the street scene. And so it went, back and forth, as each of us selected what we wanted. The air around us felt ignited, just as it had in the antique shops my father and I had once frequented. It wasn't that the paintings were so enormously valuable, but they were echoes, reminders, and this drove us on.

When we'd finished, I moved my three paintings over to my designated corner of the living room, where I'd also placed the most valuable item in the house—a Martin guitar signed by Eric Clapton, bequeathed to Elisa in the postnuptial agreement. The guitar leaned against a chair back, its leather case open on the floor, its caramel-colored surface gleaming like polished amber.

My brothers and I walked out of my father's front door for the last time, pausing on the stone steps. The shipping company had been informed of where the contents of the house would be transported. Twenty-two Martin and Gibson guitars were headed to Gruhn's in Nashville to be sold on consignment, ninety-seven antique guns were being trucked to Bonhams and Butterfields in San Francisco for sale at their next auction, forty Leica cameras were being shipped to Tamarkin Camera in New York, and so on.

With General Motors prepared to declare bankruptcy any day now, the suburban Detroit real estate market was at a record low. Ford and Chrysler, too, were struggling. In just four years, the value of my father's house had dropped some 70 percent. Nor was it just Michigan that was hurting; in the aftermath of the Lehman Brothers debacle, we couldn't have chosen a worse time to sell the house and its contents.

Within a few weeks, we would also come to understand that our father's firearms collection was peppered with counterfeits, the "million dollar" pipe collection was worth all of $50,000, and the guitars and cameras were worth considerably less than what he'd paid.

"I never liked this house," Whitney said, stamping out his cigarette on the driveway. "Dad was never happy here, that's for sure."

Whitney and Bobby stood under the eaves of the house on either side of the front door, as if keeping guard. "Well, Dad lost everything while in this house," I said. "And what did he have to look forward to?"

Bobby stepped onto the driveway. My father's favorite son, he'd always remained the most detached from the events that had led us to where we now stood. "It's the end of an era," he agreed. "Let's just hope there will be some value in all of this when the dust finally settles."

I pictured the house and its contents vanishing into a cloud of dust, fifty years of accumulation vaporized in an instant. Somehow the image made me feel even lighter. Soon I would be in the clear, taking a child's tentative first steps away from her parent's outstretched arms, the joy of walking itself my true legacy.

The three of us drove in silence to River Place for our final meeting with Bill Penner, the January roads lined with soot-encrusted drifts, Detroit's own Jack White thrashing on the radio, the way Iggy Pop and the Stooges once had, or the MC5. It was an event we'd been dreading for years, the reading of the will, although given that the company would soon be dissolved anyway, there was less to worry about; the Stroh Companies, Inc. shares now worthless, the only wild cards were our father's settlement agreements with Elisa and my mother. With any luck, his house and the cash he'd left in a bank account would cover those liabilities. Whitney and I were in the midst of our own divorces, and Bobby had just married his third wife. All of us were knee-deep in legal documents already, trying to find new beginnings.

We drove through a residential Detroit neighborhood that looked like a checkerboard with missing squares; whole blocks of abandoned houses had been leveled, while the remaining, occupied houses stood alone like homesteads in *Little House on the Prairie*, each house sometimes an entire block from its next-door neighbor. With a population drop of over 50 percent, the city was returning the valueless land to farmland, trying to consolidate the occupied area into a more sustainable footprint. No longer could Detroit afford the trash collection, police force, and fire protection in so widespread an area, and the grocery chains had fled the city because of the ever-rising crime, putting vegetables in high demand. Come springtime, grassroots urban renewal groups would be working the fields, planting everything from romaine to rutabaga.

"Just surreal," I said, staring out the window. "I mean . . . I can't believe we still own an office building down here."

Bobby kept his eyes on the road. "John's trying to get government leases now." He turned down the volume on the radio. "But the city can't even afford the infrastructure upgrade to keep the stoplights running—Did you know they're trying to sell the entire stoplight grid to a private contractor?"

I knew our delayed response to Detroit's downfall was born of attachment, a resistance to change that was similar to my grandfather's refusal to water down the beer formula after the war, and to Uncle Peter's late entry into the light-beer market. All this principled resistance, while in some ways admirable, had ultimately led to the family's unraveling. Much like the automobile manufacturers here, we simply weren't nimble enough; we'd waited too long even to close our flagship brewery. And so, twenty-four years later, still stubbornly anchored in the city where we'd made our name, we were subject to the vicissitudes of every economic contraction and political scandal. Our recently resigned mayor, Kwame Kilpatrick—convicted of obstruction of justice, assault of a police officer, racketeering, tax evasion, extortion, and mail fraud—had been the final leveling blow.

Bobby turned up the car heater. We passed more snow-covered fields mixed in with houses, some of them torched. Burning down abandoned dwellings was entertainment in Detroit, like going to the movies. The arsonists would barbecue over the embers, swigging forty-ounce bottles of St. Ides or Schlitz Malt Liquor while sirens raged through the night.

Bobby turned right onto Jefferson Avenue, and heading toward downtown, we passed the site where Uniroyal had stood, its steel-lined walls built during World War II to withstand aerial bombardment. Once it had been a place of wonder as much as danger, the badlands of Detroit, capturing my imagination as a young artist. I'd wanted to make something lasting out of all this waste; I still did. Only I needed an ending, I felt, before I could begin. Perhaps this was it.

Lately I'd been reading articles about other artists who were finding inspiration in Detroit. "I hear a lot of artists are moving here," I told Bobby and Whitney.

"Can't argue with the rental rates," said Bobby.

"I hear the DIA is being renovated," Whitney chimed in from the backseat.

"And expanded," I said. "They're spending, like, a hundred and fifty-eight million or something."

Funded in its heyday by automotive and newspaper money, the Detroit Institute of Arts housed one of the most expansive collections in the world, ranging from ancient Egyptian works to contemporary art. The new Museum of Contemporary Art Detroit had recently been written up in the *New York Times*, putting Detroit back on the map, artwise. The city seemed overrun with artists setting up studios in abandoned buildings, showing their work in makeshift galleries, and sipping fair-trade coffee in what looked like domestic war zones. They, too, believed in the redemptive power of danger, these gutsy people, forging a life with no certainties—the kind of life Bernhard Stroh had opted for back in 1850, when the water in Detroit tasted so fresh.

Turning left onto Joseph Campau, Bobby continued down to Stroh River Place and parked next to the river in the same lot where Coleman Young and Uncle Peter had once made their deal. The fact that we still owned and parked in this lot was proof that things had not gone as planned along the riverfront. In every other American city, waterfront lots were prime real estate developed into high-end residential and commercial areas. Not so in Detroit. Our waterfront parcel had actually plunged in value since that fateful day, twenty-five years ago, when those two city fathers shook hands on a future that was never to be.

I looked around at the handful of parked cars studding our acreage of buckling concrete, trying to picture a field of corn in its place. The idea was kind of hopeful. Anybody could see it was time to start over.

ACKNOWLEDGMENTS

Many people have unerringly supported me in the writing and publishing of this book. My agent, Rob McQuilkin, is the greatest champion a writer could ever hope for. I am very grateful to my editor at Harper, Jennifer Barth, for her keen eye and bright inquisitiveness, and for her belief in this book; I thank Zoe Rosenfeld for always seeing the forest for the trees and, without fail, the outline of each tree, in the early editing work we did together.

My early readers nurtured this project at its most critical stages: Maria Massie, Katie Fleischer, Claire Sanders Swift, Tony Meier, Marnie Burke de Guzman, Alan Black, Heather Cappiello, Rachel Howard, Pam Bohner, D'Arcy McGrath, Dave Dederer, Elsa Dixon, Lindsey Crittenden, Audrey Ferber, Monica Wesolowska, and Arkady Shirin. I am deeply grateful for your support, and for your friendship.

I feel immense gratitude for my teachers, Tom Barbash and Julie Orringer, for initiating the spark; for my late father

Eric Stroh and late brother Charlie Stroh, whose lives left deep welts, only to open channels much deeper yet; for my living brothers, Bobby Stroh and Whitney Stroh, whose early support came with characteristic humor and grace; for my mother, Gail Marentette, who not only warmed to the idea but embraced it with all her magnificence; for the rest of the Stroh family, whose tolerance knows no bounds; for Arkady Shirin, who shouted from the mountaintops that I could—and would—write this book; and most of all, for Mishka Shirin-Stroh, my son and great inspiration, who made such colossal sacrifices along the way.

ABOUT THE AUTHOR

Frances Stroh was born in Detroit and raised in Grosse Pointe, Michigan. She received her BA from Duke University and her MA from Chelsea College of Arts in London as a Fulbright Scholar. She practiced as an installation artist, exhibiting in Los Angeles, San Francisco, and London, before turning to writing. She lives in San Fransisco, California.